Mobile Device Security
FOR
DUMMIES®

by Rich Campagna, Subbu Iyer, and Ashwin Krishnan

Foreword by Mark Bauhaus
Executive Vice President, Device and Network Systems Business Group, Juniper Networks

WILEY

John Wiley & Sons, Inc.

Mobile Device Security For Dummies®

Published by
Wiley Publishing, Inc.
111 River Street
Hoboken, NJ 07030-5774

www.wiley.com

Copyright © 2011 by Wiley Publishing, Inc., Indianapolis, Indiana

Published by Wiley Publishing, Inc., Indianapolis, Indiana

Published simultaneously in Canada

For general information on our other products and services, please contact our Customer Care Department within the U.S. at 877-762-2974, outside the U.S. at 317-572-3993, or fax 317-572-4002.

For technical support, please visit www.wiley.com/techsupport.

Wiley also publishes its books in a variety of electronic formats and by print-on-demand. Not all content that is available in standard print versions of this book may appear or be packaged in all book formats. If you have purchased a version of this book that did not include media that is referenced by or accompanies a standard print version, you may request this media by visiting http://booksupport.wiley.com. For more information about Wiley products, visit us www.wiley.com.

Library of Congress Control Number: 2011932276

ISBN 978-0-470-92753-3 (pbk); ISBN 978-1-118-09379-5 (ebk); ISBN 978-1-118-09380-1 (ebk); ISBN 978-1-118-09399-3 (ebk)

Manufactured in the United States of America

10 9 8 7 6 5 4 3 2 1

WILEY

About the Authors

Rich Campagna is a Director of Product Management at Juniper Networks. His team is responsible for defining product strategy for Juniper Networks' Junos Pulse Business Unit, including the Junos Pulse Mobile Security Suite, the SA Series SSL VPN product family, Juniper's Unified Access Control product family, the Junos Pulse Application Acceleration product family, and the Junos Pulse client software. Rich was a co-author for *Network Access Control For Dummies*. Prior to joining Juniper Networks, Rich was a Sales Engineer at Sprint Corp. He received an MBA from UCLA Anderson School of Management and a BS in Electrical Engineering from Pennsylvania State University.

Subbu Iyer is a Senior Product Manager at Juniper Networks. He drives the product strategy of the Junos Pulse product line, which provides a variety of integrated network services on desktops and mobile devices, including smartphones and tablets. His prior experience includes over eight years at Cisco where he held various senior architecture and engineering roles focusing on application-aware networking, security, and WAN acceleration. He has extensive experience in software development and marketing of products in the areas of Application and Network Security, including remote and LAN access control. Subbu holds an M.S. in Computer Engineering from the University of Arizona, Tucson and an M.B.A. from the Haas School of Business, UC Berkeley.

Ashwin Krishnan is a Director of Product Management at Juniper Networks, where he runs the product management team that is responsible for the high-end SRX product line (which has the leading market share position according to Infonetics Research) and SRX service provider business. He also heads up a cross-functional mobile security team that is focused on defining the strategy and solutions for infrastructure and services protection in the mobile network. His prior experience includes over five years at Nokia where he held various senior product management, architecture, and engineering management roles focusing on core infrastructure, service control, and intelligent subscriber gateway products. Prior to that he has held various lead technical roles at 3Com, Octel, Hughes, and Wipro. He is a frequent speaker at security and mobile conferences (NGMN, 4G world, Informa, and so on) and regularly blogs about all aspects of security. He has over 17 years of industry experience with specialization in wireless, security, and IP networking. He attained his Bachelor of Science degree from the National Institute of Technology, Warangal, India in 1991.

Dedication

Rich Campagna: To Brooke — Daddy loves you!

Authors' Acknowledgments

Subbu Iyler: I would like to thank my wife Manju, and daughter Anoushka, for their constant motivation, encouragement, and support throughout the writing of this book.

Ashwin Krishnan: I would like to thank Radhika, my wife; Ananya, my daughter; and Jackie, our dog for supporting me through the process of creating this book while I was ostensibly doing chores to write it, including walking the dog (sorry Jackie). Thanks for putting up with my vagaries. And to my mom, Indira Ananthakrishnan, who is a renowned author herself, for instilling in me some of your book writing genes.

To the "numero uno" team at Wiley for providing excellent feedback throughout the process and helping get the book into its final finished form.

And finally to our Juniper in-house editor-in-chief, Patrick Ames, who helped instigate the idea of writing the book and cajoled, threatened, and pleaded with us throughout the course — without you this book never would have happened.

Publisher's Acknowledgments

We're proud of this book; please send us your comments at http://dummies.custhelp.com. For other comments, please contact our Customer Care Department within the U.S. at 877-762-2974, outside the U.S. at 317-572-3993, or fax 317-572-4002.

Some of the people who helped bring this book to market include the following:

Acquisitions and Editorial

Project Editor: Kim Darosett

Senior Acquisitions Editor: Katie Mohr

Copy Editor: Heidi Unger

Technical Editor: Rob Cameron

Editorial Manager: Leah Cameron

Editorial Assistant: Amanda Graham

Sr. Editorial Assistant: Cherie Case

Cartoons: Rich Tennant
(www.the5thwave.com)

Composition Services

Project Coordinator: Sheree Montgomery

Layout and Graphics: Timothy C. Detrick, Nikki Gately, Corrie Socolovitch

Proofreaders: Context Editorial Services, John Greenough

Indexer: Broccoli Information Management

Special help: Colleen Totz Diamond, Kimberly Holtman

Publishing and Editorial for Technology Dummies

Richard Swadley, Vice President and Executive Group Publisher

Andy Cummings, Vice President and Publisher

Mary Bednarek, Executive Acquisitions Director

Mary C. Corder, Editorial Director

Publishing for Consumer Dummies

Kathleen Nebenhaus, Vice President and Executive Publisher

Composition Services

Debbie Stailey, Director of Composition Services

Contents at a Glance

Table of Contents

Foreword

. . . .

*T*he sweep of mobile devices into our lives has transformed business IT in only a few short years. If you are holding this book in your hands, then you have no doubt encountered this massive change firsthand, and you are looking for answers. Looking out over the next few years, mobile devices will continue to transform the way that we do business.

New form factors such as tablets and "dockable" smartphones will allow users to replace laptops and desktops, DVRs, radios, DVD players, and many other "fixed" devices — untethering us completely. Ubiquitous network access through Wi-Fi, 3G, 4G/LTE, and beyond will allow us to do work anywhere, and at any time. Advances in peripherals and device-to-device interaction will integrate these devices into our lives much more seamlessly, no longer requiring us to remove them from our pockets.

All of this freedom, however, brings forth a huge challenge for corporate IT. It wasn't that long ago when our IT departments required that users access corporate data and applications from a specific, corporate-issued device (typically a RIM BlackBerry).

Starting with the release of the Apple iPhone, however, users began to demand choice, ushering in a new era of consumerization that will forever change enterprise IT. Today's users typically purchase their own mobile device of choice, and they find a way to connect it to the corporate network. Your challenge is to provide the flexibility that your users need, without sacrificing security, and this book, *Mobile Device Security For Dummies,* provides a complete look at the best practices for allowing you to meet that challenge. The authors are at the forefront of mobile device security strategy and product development, and are particularly well-suited to provide a balanced view of the current state of security concerns, and recommend ways to assuage those concerns. They regularly advise a range of customers across just about every domain and industry vertical, so chances are they have experience dealing with other organizations just like yours, with similar challenges.

The mobile world is evolving quickly — with new devices, operating systems, capabilities, and even threats emerging with every passing day. To meet the inexorable mobile changes, you'll want best practices on how to manage these challenges while adapting the new truths of mobility in the enterprise. So go ahead: Ride the mobile wave safely with the best tools and practices available in the market.

Mark Bauhaus

Executive Vice President

Device and Network Systems Business Group

Juniper Networks

Introduction

● ●

Mobile devices, including smartphones and tablets, rule the market-place. Regardless of whether these devices are employees' personal devices or company-issued, you need to adopt best practices in an effort to secure them. It's an effort, because very little planning and budget are devoted to these powerful little devices; but you have to have a plan for securing your company and its network, people, resources, and information.

This book helps you plan for mobile device security in your business and extend it into the lives and homes of your company's employees. Having a plan helps you plead your case to management, and this book gives you the background you need to make the best decisions for your own implementation of mobile security, management, and control.

We (the authors) work on mobile security software and hardware and have worked for many years on security software implementation throughout the world. This is not to emphasize our massive intelligence in the matter, but rather point out that we've seen just about every marketplace and every issue that various IT departments and network administrators face in imple-menting a mobile strategy. And because we work for Juniper Networks, on the Junos Pulse product team, we know intimately what our customers need. In this book, we give you a view of the mobile security world from a collec-tive viewpoint: beginner, implementer, and successful provider. Regardless of whether you choose Junos Pulse or another solution, or implement your own customized solution, this book helps you understand the threats facing mobile device adoption today and implement the current best practices for securing these devices in the enterprise (the best practices we've learned the hard way).

About This Book

This book isn't meant to be read from cover to cover. It's more like a refer-ence than a suspense novel. Each chapter is divided into sections, each of which has self-contained information about a specific task in setting up a mobile device security solution.

You don't have to memorize anything in this book. The information here is what you need to know to complete the task at hand. Wherever we mention a new term or are possessed by the need to get geeky with the technical descriptions, we've been sure to let you know so that you can decide whether to read or ignore them. Aren't we thoughtful? You're welcome.

Mobile device security has several players: you, the administrator; the mobile device users; management, who must fund security solutions; vendors, who create and sell their solutions; and a shifting crowd of nefarious hackers, thieves, and competitors who are looking for cracks in your wall. While you might find other books about mobile device security, you won't find one that makes you aware of all the players all the time. This is a new-school book about new-school technology.

Foolish Assumptions

We make a few assumptions about who you are. For example, we assume you bought this book to learn more about mobile device security in the enterprise, hence we assume your job is as an enterprise IT or network administrator. If you're not one of those industrious people, we assume you might be in IT management or even sales management. In short, you work for a company whose employees all connect to the network with their mobile devices, and you're supposed to be, somehow, one of the people who control this.

We have bad news and good news for you. The bad news is that we're sorry you are in this position. If you haven't had security problems yet, you will. We've seen many customers seeking security solutions in our lifetimes, and the good news is that this book details the threats facing mobile device adoption today and the best practices that you can implement for securing them in the enterprise.

Conventions Used in This Book

We know that doing something the same way over and over again can be boring (like Mr. Rogers always wearing the same kind of sweater), but sometimes consistency is a good thing. In this book, those consistent elements are called *conventions*. In fact, we use italics to identify and define new terms you might not recognize, just like we've done with the word *conventions.* Additionally, when we type URLs (web addresses) within a paragraph, they look like this: www.wiley.com.

That said, throughout this book we use the terms *smartphones* and *mobile devices* interchangeably. Sometimes only smartphones have the capability of over-the-air transmission, but new mobile devices are coming that could far surpass even the smartphone's capabilities. So we use *smartphone, mobile device, iPad, iPhone, Android, BlackBerry,* and other terms interchangeably, too.

At the end of many chapters, we include a case study based on experience we've gained from our customers who have grappled with similar situations. It's the only way we can justify how many miles we've flown during the past five years, but more importantly, we hope you can benefit from this running example of how you might implement some of the policies we discuss throughout the book.

That's about it. Mobile device security is so new that the only convention you share with everyone else around you is a feeling that your data isn't secure. At all. But fear not — it will be after you implement the policies discussed in this book.

How This Book Is Organized

This book is organized into five main parts. Don't feel that you need to read these parts in sequential order; you can jump around as much as you like, and each part is meant to stand on its own.

Part 1: Living Securely in the Smart World

Sometimes it's comforting for authors to describe the world you live in. Part I of this book describes the world that you're trying control. You'll be able to find yourself here, in one of the chapters, in one of the scenarios. Misery loves company, and eventually by Chapter 3, we ask you to stop fighting the hordes of mobile devices in your environment and instead embrace them. Embrace, adapt, protect, and manage are the four stages of living securely in this smart new world.

Part II: Implementing Enterprise Mobile Security

Part II assumes you've given up the "no mobile devices permitted onsite" fight and taken down the signs. Implementation starts by creating policies and then managing and monitoring them. It's not rocket science, and chances are you already do many of them today. This part helps you put your policies together and perform the real trick: Make your mobile device policies conform to existing compliance policies so you don't have to redo policies for the whole company.

Part III: Securing Smart Device Access

Part III moves from the policy to the real world — your network. How do you build the system of monitoring, accepting/rejecting, or limiting access to the hordes of devices entering your main, branch, and remote offices? Not to reveal the ending too much, but you're going to leverage technology to provide granular, application access control.

Part IV: Securing Each Smart Device

At some point, you have to touch your customer. It's time to roll out the policy, programs, and technology to encrypt, protect, and back up the device hoards. You don't want to be in upper management, anyway.

Part V: The Part of Tens

Indispensable places and checklists tend to come in lists of tens, and mobile device security is no different. Turn here often as you read the book, and come back when you're done.

Icons Used in This Book

To make your experience with the book easier, we use various icons in the margins of the book to indicate particular points of interest.

Whenever we give you a hint or a tip that makes an aspect of mobile device security easier to understand or speeds the process along, we mark it with this little Tip thingamabob. It's our way of sharing what we've figured out the hard way so you don't have to.

This icon is a friendly reminder or a marker for something that you want to make sure that you keep in mind, or remember, as the icon says.

Ouch! This icon is the equivalent of an exclamation point. Warnings give you important directions to prevent you from experiencing any nightmares. (Well, at least where security is concerned. Offering premonitions about your personal life costs extra.)

Sometimes we feel obligated or perhaps obsessed with some technical aspect of mobile security. We are geeky guys, but mark this info thusly so that you know it's just geeky background information.

Where to Go from Here

Now you're ready to use this book. The beginning introduces basic security concepts so you're familiar with both the terminology and the state of affairs in today's mobile device security marketplace. If you're new to mobile device security, start here, or depending on your background, you may want to start by jumping straight to the meat of the discussion in Part II. Once you zoom in to what interests you, we highly recommend going to the other parts or chapters because there are key concepts and usage cases in each chapter.

If you have a mobile device on your desk right now, we recommend muting the ringer and alarms and putting it to sleep for awhile. These devices don't like to be corralled at first, and if they see you reading this book, they'll start acting strange for an hour or so.

If you ever want to see what we authors really do, and some of the products we actually get paid to work on, check out Junos Pulse at the Juniper Networks website, www.juniper.net/pulse.

Part I
Living Securely in the Smart World

The 5th Wave   By Rich Tennant

"Don't be silly - of course my passwords are safe.
I keep them written on my window, but then I
pull the shade if anyone walks in the room."

In this part . . .

By the end of reading Chapters 1 and 2, you will recognize that your best option for securing your corporate network is to embrace the hordes of mobile devices on your campus — well, *embrace* may be going overboard, but at least you should acknowledge their existence. You can't live with mobile devices, and you can't live without them. What's the answer? Embrace, adapt, protect, and manage. Those are the four stages of living securely in this smart new world — and the message of Chapter 3.

Chapter 1

What's So Smart About a Phone, Anyway?

*T*he late 2000s and early 2010s ushered in a new era of mobility in the enterprise. Prior to this time, truly productive mobility required users to have a laptop, a mobile phone, and possibly a personal digital assistant (PDA) in order to be as productive offsite as they would be at the office. The rise of the smartphone, however, has changed all of that. Now users can get as much done with a device that fits in their pocket as they could when three separate devices were required to accomplish the same tasks. With tablets reaching widespread adoption as well, many users and organizations are trading in their laptops and desktops and replacing them with these new devices.

Your enterprise may have worked for years on strategies for the use of Microsoft Windows (on laptops and desktops) and the Research In Motion (RIM) BlackBerry OS (on smartphones). In addition to the tools that Microsoft and RIM provide to manage, update, and secure these operating systems, your enterprise may have invested in a number of third-party components to help secure these systems further.

However, with the overwhelming demand to bring smartphones and tablets into the enterprise, many IT departments are forced to allow these devices into their networks, in many cases without properly adopting security policies and procedures and without rolling out the appropriate solutions to secure these devices. Because you have picked up this book, you are most likely concerned about how to successfully adapt what you know about security to an extremely wide range of mobile devices.

In this chapter, we describe the various mobile device *form factors* (the physical dimensions of the devices), the operating systems that run on those devices, and the types of data connections you need to be concerned with when planning a mobile device strategy. We also explain how the applications and data running on these devices will impact your mobile device security strategy.

Additionally, this chapter gives you an overview of the many considerations that you need to take into account when you decide to allow mobile devices to connect to your corporate network. We give you an introduction to the components that make up a successful mobile device security deployment, and then the rest of the book goes into the details.

Finally, the chapter ends with an introduction to a case study of AcmeGizmo, a fictional company. At the end of many chapters of the book, you'll find case study excerpts to show how this example company chose to deploy various security products and solutions to secure its employee smartphone deployment.

Exploring Different Mobile Devices

The many different mobile computing devices available in the market today range in sizes small enough to fit in your pocket to large enough to require a backpack or over-the-shoulder bag. In this section, we introduce the major form factors of mobile computing devices.

Smartphones and tablets

Smartphones and tablets fuel today's mobile device explosion. Tens of millions of these devices have been adopted in the last few years, with forecasts of tens of millions more to hit the market in the near future. These devices have very quickly found their way into the enterprise, and they're the primary subject of this book. Many of these devices (and their associated operating systems) were designed for the consumer market, and vendors have added more enterprise-friendly functionality over time. Still, their roots as consumer mobile devices have left some enterprises dissatisfied with or unsure of the risk level of these devices.

Typically, these devices run operating systems specifically designed for smartphones: primarily, Apple iOS, Google Android, RIM BlackBerry OS, Microsoft Windows Mobile (up to version 6.5) and Windows Phone (version

7.0+), and Nokia Symbian (which Nokia is in the process of abandoning in favor of Microsoft Windows Phone 7), though there are several other operating systems on the market today.

Smartphones

The line between a smartphone and a traditional feature phone blurs with each new generation of devices on the market. Vendors continually add more and more functionality to traditional feature phones, while at the same time, lower-end smartphones are introduced to the market in an effort to appeal to the more price-conscious consumer.

That said, there are still distinctions between the typical feature phone and a smartphone. Smartphones are frequently described as handheld computers. All have built-in mobile phone functionality, but what differentiates a smartphone from a traditional mobile phone is the ability for the user to install and run advanced applications (in addition to the ability for independent developers to actually build and distribute those applications). It is this ability to add third-party software that makes smartphones an incredible productivity tool for enterprise users, while at the same time makes them susceptible to malware and other types of attacks targeted at those systems. This book helps you to balance productivity gains with security as you enable your end users to use these advanced devices.

In recent years, many smartphones have transitioned to the touchscreen interface, as shown in Figure 1-1, though some still feature a stylus as an input device. Some smartphones include a physical keyboard; others do not. Increasingly, smartphones feature large screens and powerful memory and processors.

One of the big appeals of smartphones today is the availability of third-party applications, typically through application stores or marketplaces, such as iTunes (from Apple), Ovi (from Nokia for Symbian devices), or Android Market (from Google). These marketplaces are where users typically go to purchase and download applications.

In recent years, many enterprise applications have started to make their way into these marketplaces, enabling employees to easily acquire software that helps them to be more effective and productive in their jobs. One of the most common examples is the killer application: e-mail or, more generically, messaging. E-mail is almost always the first application used by enterprises on mobile devices. As enterprises have embraced these mobile devices more completely, they have moved on to more comprehensive business applications such as online tools access, database applications, and sales force applications such as Customer Relationship Management Software (CRM). In fact, you would be hard pressed to find a type of application that hasn't been ported to mobile devices somewhere.

Figure 1-1:
Both the
iPhone (left)
and Droid
(right) sport
touchscreen
interfaces.

Tablets

Tablets are most commonly identified by their slate shape (see Figure 1-2). They use touchscreens as their primary input device. You'll find a wide variety of devices in this style, but today's devices generally run either a version of Microsoft Windows or one of the smartphone operating systems. Tablets running smartphone operating systems such as Apple iOS or Google Android are among the most popular tablets on the market today.

In this book, we focus on tablet devices that run one of the smartphone operating systems. Devices running one of the several Windows variants can be treated very much like a laptop or a netbook from a security perspective, because they are capable of leveraging the endpoint security and desktop management tools available for those other devices running Windows. As a result, devices that run full versions of the Microsoft Windows operating system are outside of the scope of this book. Devices running the Microsoft Windows Phone or Windows Mobile operating systems, by contrast, are covered in detail in this book.

Devices such as Apple's iPad (which runs iOS), or one of the many Google Android-based tablets on the market, are similar to smartphones in terms of their capabilities and the security issues that the typical enterprise should be concerned about when allowing these types of devices to access corporate networks. Because these devices run the same operating systems as their smartphone brethren, the security implications and the security policies applied to each are exactly the same.

Figure 1-2:
The iPad
is a type of
tablet.

Laptops and netbooks

Notebooks (or *laptops*) and netbooks are traditionally used as the primary computing devices in many enterprise environments for mobile users (though trends are quickly changing that positioning). Typically, these devices run versions of the major desktop operating systems: Microsoft Windows or one of several popular distributions of Linux (Red Hat, SUSE, Debian, Ubuntu, and so on). Macintosh laptops generally run a version of Mac OS X. Notebook devices are most often based on x86 processing and come in a variety of sizes, with varying hard disk, memory, and other components.

Notebooks have been around in the enterprise for a very long time, and most IT departments have made significant investments in securing and patching these devices. This book does not emphasize or discuss security strategies for these types of devices, and you can easily find a variety of resources and industry knowledge on how to securely deploy these types of devices for enterprise use.

Netbooks are smaller and less powerful than laptops. These devices are specifically built for the low-end consumer market and have not seen wide-spread adoption in the enterprise, though you may encounter end users who wish to leverage these devices to access the corporate network as personal devices for use when working from home or when traveling. Netbooks typically run scaled-down versions of Microsoft Windows or Linux operating systems, which do not significantly alter the security risk of the devices, and the devices should be secured in a similar fashion to those machines running full versions of the operating system (despite the fact that they have less functionality to exploit).

Aside from notebooks and netbooks, there are other device types on the market, such as the tablet PC, though these devices have never gained widespread popularity and are quickly being phased out in favor of tablets running operating systems designed for smartphones and tablets (such as Apple iOS and Google Android). For this reason, we don't cover these devices in detail in this section or in this book.

Other computing devices

There are a variety of other computing devices that are probably attached to your corporate network, but as with laptops and netbooks, these devices are outside the scope of this book. Some of these devices include desktop PCs, feature phones, and warehouse and inventory devices.

Examining Operating Systems for Mobile Devices

So many systems, so little time. With so much overlap and so little difference between many of the device types discussed in the preceding section, it can be confusing to tell just by looking at a device what security mechanisms should be applied to it. It's important to think about the operating system running on the device because that has a big impact on the type and availability of security products that should be applied to the device.

The operating system is the primary interface between the underlying hardware and the applications running on the device. Among other things, the operating system provides a (mostly) generic mechanism for application developers to write a single application and run it on multiple hardware devices running the same operating system. For this reason, the operating system is the primary distinction that we use in this book to differentiate between mobile devices (the primary subject of this book) and everything else.

A large number of mobile operating systems are available on the market today. Only a few of these have really taken off to the point where you are likely to see large numbers of users adopting them for use in the enterprise. Most vendors provide support for, at most, the top five or six operating systems on the market. You will also find that these five or six operating systems represent the overwhelming majority of phones, so that is not likely to become a significant problem. Security vendors also keep a close eye on the market for mobile operating systems, however, and as new operating systems gain or lose market share, you might see coverage change with newer versions of the security software that your organization has adopted.

 You have the option of either allowing all devices onto your network or restricting access to a smaller number of devices. We recommend that you restrict access only to those operating systems that you feel comfortable being able to secure, so that you do not put your organization's sensitive corporate data at risk.

The following sections briefly describe the major operating systems in use on mobile computing hardware and also highlight which operating systems fall under the mobile device security strategies discussed in this book.

Apple iOS

Apple's iOS runs on a range of devices, including the iPhone, iPad, iPod Touch, and Apple TV. Apple tightly controls the operating system and does not allow it to be used on third-party hardware, so it is found only on Apple hardware devices. iOS (running on iPhone) is commonly known as the operating system that really started the current mobile/smartphone revolution in the enterprise. Prior to the iPhone, RIM's BlackBerry devices were the de facto standard in the enterprise, but massive consumer adoption and employee demand for corporate access from the iPhone changed that, forcing many enterprises to adopt new mobile device strategies.

iOS is based on Mac OS X, Apple's desktop and laptop operating system. As with other mobile operating systems, iOS includes a software developer kit (SDK) that allows third-party developers to write and distribute applications for iOS devices. Applications for iOS are published through Apple's App Store, which includes hundreds of thousands of downloadable applications.

Apple's tight control of both its hardware and the applications installed on the iOS operating system is both a good thing and a bad thing from a security perspective. On the plus side, the tight control of applications allows Apple to screen applications for (among other things) security prior to distribution. The hardware control allows Apple to lock down its operating system software, exposing fewer functions that might potentially be exploited.

On the downside, Apple has prohibited many third-party security applications, such as antivirus software, from being made available on the iOS platform, taking some of the control over security from the hands of the enterprise IT administrator. Thus far, Apple has done a good job of keeping malware and viruses from making their way to the App Store, so that hasn't become a huge issue.

Key security distinctions: iOS versus Android

Apple iOS and Android are the two most talked about (and adopted) smartphone/tablet operating systems on the market. Both have gained widespread popularity with mobile application developers, with hundreds of thousands of applications available for each platform through various application marketplaces. There are a couple of key distinctions between iOS and Android, however, that are important to point out. These differences are important because they have significant security implications and make it that much more important to carefully plan your security deployment for Android devices.

Here are the main differences between Android and iOS:

✔ **Malicious applications.** Apple tightly controls and reviews every application before allowing it to be posted to its App Store. This (according to Apple) helps to mitigate the chance that malicious applications can find their way onto devices running iOS. As an open source project, however, Android's developer community is self-policing. This means that any application developer can post an application, and it is up to the community to determine whether an application is malicious in any way, and lobby to have it removed. As a result, a number of potentially malicious applications that target Android devices have been found only after end users downloaded and used them.

✔ **The Apple App Store.** Apple's iOS offers only the one App Store, from which users can download applications to their devices. (In 2010, Apple began offering new Application Programming Interfaces [APIs] that allow enterprises to develop their own application stores, which allows enterprises to publish and distribute their own applications directly to their employees. APIs are a set of specifications and interfaces that allow an application to communicate with the underlying operating system.)

With Android, however, there are a number of places from which end users can download applications. Google's Android Market is the primary app store for Android devices and comes installed on most devices running Android. There are, however, many other application stores that can be leveraged by Android devices, many of which are less heavily policed, are known for distributing cracked/hacked software, and represent a big security concern for Android devices accessing corporate networks. It might be a good idea to prohibit your Android users from accessing any of these other application stores.

It is a good idea to train your end users to exercise caution when downloading applications, even from the Android Market itself. Users should download only from trusted sources, and should read reviews to ensure that the applications that they are downloading aren't already causing other folks issues. Android developers and users do attempt to police the marketplace, notifying Google as soon as possible if malware is present; and thus far, the window of exposure for Android malware has been minimal, but still very real nonetheless.

✔ **Operating system fragmentation.** Operating system fragmentation is an issue to be aware of on Android devices. With Apple iOS, every device is capable of running the same versions of the operating system, and Apple makes it easy for users to upgrade to the latest versions of iOS through its iTunes software. With Android, however, the hardware and the software are created by two separate entities, and hardware vendors frequently make additions and modifications to the operating system before distributing it. At the same time, some device manufacturers limit or prohibit upgrades to newer versions of the operating system, potentially exposing users to security issues that have been resolved in newer versions. Specifying and controlling which versions of Android can access your network might be a prudent step toward mitigating these risks.

✔ **Sandboxing applications.** Both Android and iOS *sandbox* applications, prohibiting them from communicating with other applications on the devices. Apple has made strong statements indicating that this sort of security, along with its review of every application before it is posted to the App Store, is sufficient to keep malicious code from being distributed to iOS devices. As a result, Apple prohibits third-party endpoint security vendors from building software such as antivirus and antimalware for iOS. It remains to be seen whether this strategy will continue to scale and succeed, but as this book went to press, Apple's strategy has been successful.

It is important to note that we are not attempting to sway enterprises away from allowing users to adopt Android devices. This section is merely meant to highlight some of the additional concerns to keep in mind when allowing Android devices onto corporate networks. These issues can be mitigated or eliminated through proper security planning, policy, and the use of third-party security software.

Google Android

Google's Android operating system became extremely popular over the 2009–2011 period. While sponsored by and commonly associated with Google, Android is an open source operating system with many contributors. Android is based on Linux, as is common with several of the mobile operating systems described in this section.

As with Apple's iOS, there are hundreds of thousands of applications available for the Android platform. The Android operating system can be found on smartphones and tablets from a wide variety of handset vendors, including Motorola, Samsung, Dell, HTC, and more. In the second half of 2010, Android became the unit market share leader for smartphone operating systems in the United States.

With the Android operating system, the OS itself is open source, which means that malicious entities might have an easier time finding exploits in various versions of the OS. On the other hand, this open source nature also means that a large community of contributors are keeping an eye on the development of the OS and contributing work. The primary security concern associated with Android systems is the lack of policing on the Android marketplace, as well as the availability of non-Google sponsored marketplaces. Several well-known malware applications have now found their way onto Android systems, with more expected to come in the future. This strengthens the need for a comprehensive security story on Android devices.

RIM BlackBerry OS

Research In Motion's (RIM's) BlackBerry operating system has been wildly popular in the enterprise for a number of years. Until recently, with the newest wave of smartphones on the market, it has been the de facto standard for corporate data and application access from a mobile device. This OS became popular in the enterprise due to its native support for corporate e-mail, as well as the management and security functionality that is native to the operating system.

Key to the management and security features is the BlackBerry Enterprise Server (BES), which sits inside of the corporate network and provides authentication, security of data in transit, and security of the device itself. The built-in security does not cover everything, however, and a number of third-party security products on the market complete the BlackBerry end-to-end security story.

Many IT administrators, including some reading this book, wish they could return to the days where they needed to support only a single mobile device operating system (BlackBerry OS), which can be controlled by a single management platform (BES). Unfortunately, the "consumerization" of IT has led to the adoption of myriad other devices by corporate users, so the task of securing devices has become much more complicated (hence the need for books like this one).

Most BlackBerry phones on the market run RIM's BlackBerry OS, though it is expected to be replaced by a new OS (currently known as the BlackBerry Tablet OS; see the following section). Blackberry OS version 7 will actually be this new operating system, rather than a continuation of the prior versions of the Blackberry operating system.

RIM BlackBerry Tablet OS

BlackBerry Tablet OS is, as of early 2011, a new operating system from RIM that runs on the RIM Playbook tablet. This operating system represents a major shift for RIM, as all of its devices have run some version of the BlackBerry operating system. This new OS is based on a real-time OS, similar to Unix, known as QNX.

RIM has announced plans to transition all of its devices to this new operating system as of BlackBerry 7. Given the tremendous popularity of RIM devices in enterprise environments, it is likely that many mobile device security vendors will adapt their products to support this operating system as BlackBerry 7 devices begin to hit the market. While this operating system may not be a big concern for the corporate IT department in 2011, moving forward, it is something to plan to support.

Microsoft Windows Mobile and Windows Phone

Windows Mobile and Windows Phone are Microsoft's mobile operating systems. Until version 6.5, Microsoft's mobile device OS was known as Windows Mobile and was heavily focused toward the enterprise. Version 7 onward is known as Windows Phone, and at least initially, the operating system is built primarily for consumer use. In early 2011, Microsoft's mobile operating systems continue to fall in market share, making them far less popular than several of the other operating systems described in this section.

Because Windows Mobile (6.5 and prior) was targeted toward the enterprise, it includes many built-in security features and provides the OS capabilities and APIs for third-party security developers to create applications that help secure these platforms. Over time, Microsoft will be phasing out Windows Mobile 6.5 in favor of the newer Windows Phone 7 operating system.

The continuation of release number from 6.5 to 7 is a bit of a misnomer, because Windows Phone 7 is an entirely new operating system and is very different from Windows Mobile (6.5 and prior). A number of functions are currently missing from Windows Phone 7, including virtual private network (VPN) support and on-device encryption, and that prohibits it from being properly secured and connected to enterprise networks.

It is likely (though not confirmed) that over time, Microsoft will add some of these missing enterprise features. For the time being, however, it is important to note that Windows Phone 7 does not necessarily include features that your enterprise might be using on Windows Mobile 6.5, and that your existing security products might not support this newer operating system yet.

Nokia Symbian

Symbian is an open source operating system managed and maintained by Nokia (though it is licensed by the nonprofit Symbian Foundation). Symbian is primarily found on Nokia devices today, with prior licensees, such as Sony Ericsson, Samsung, and others transitioning to competitive platforms, such as Android. Even Nokia's dedication to the platform is questionable, as it has begun to introduce new high-end smartphones based on the MeeGo operating system. Nonetheless, despite the onslaught of new competitors over the past few years, Symbian remains the global market share leader for smartphone sales and installed base.

Having been on the market for several years, there are a wide variety of security solutions available for Symbian. This is important because there have been several outbreaks of malicious code/applications on Symbian platforms over the past several years.

In 2011, Nokia made several high-profile announcements indicating that its new devices will transition from Symbian to Windows Phone as the primary smartphone operating system. As a result, the introduction of new Symbian devices into the market will be limited. Despite this news, Symbian remains a popular platform globally and must be a part of any global organization's mobile device security strategy until the installed base of Symbian devices declines to insignificant levels as users move to newer devices.

HP Palm webOS

webOS is another Linux-based mobile operating system. After a long history of personal digital assistants (PDAs) running the Palm OS platform, Palm introduced webOS as its next-generation operating system in 2009. The Palm Pre and the Palm Pixi are the most well-known device families to run on the webOS operating system.

Despite great reviews, the webOS platform struggled to gain traction in the marketplace, especially against prominent competitors such as Apple's iOS and Google's Android. Hewlett Packard (HP) purchased Palm in 2010. HP webOS (as it is now known) is on version 2.0 as of early 2011, and HP's long-range plans for this operating system are not yet known, though a device known as the Pre 2 is on the market in several areas of the world.

As with some of the other less popular platforms, few mobile security platforms support the webOS operating system, but it is not impossible to find them, as webOS has enjoyed some success in the market, albeit small.

MeeGo

Like many mobile device operating systems, MeeGo is Linux-based. In fact, MeeGo is an open source operating system and is part of the Linux Foundation. MeeGo is capable of running on a wide range of devices, leveraging a common platform foundation with user interfaces built specifically for different types of devices.

MeeGo was first announced in early 2010 and had yet to gain significant traction with device vendors or with end users when this book went to press, though Nokia is expected to launch one or more MeeGo devices soon.

Very few (if any) mobile device security platforms currently support MeeGo, so keep that in mind if you start to hear demands from end users to support this platform. It is likely that if it increases in popularity, security vendors will adapt their products to this platform.

Samsung bada

Yet another Linux-based platform, bada is developed by Samsung. Samsung's aim is to use this operating system to replace the operating systems on both its smartphones and its feature phones, further blurring the line between the two types of devices. When this book published, Samsung had shipped only a single bada device, the Samsung Wave smartphone.

The future of bada is unclear, as Samsung also ships a variety of devices running the very popular Google Android platform. It will be difficult to find security platforms on the market that cover the Samsung bada platform, but as with MeeGo, it is likely that if this platform takes off and becomes popular, vendors will respond with products designed for or adapted to the bada platform.

Discovering Data Connections

It is no longer uncommon for a mobile device to have the ability to connect to multiple types of data networks. At the same time, it is increasingly common for sensitive corporate data to be stored directly on these devices. That means that your security deployment needs to have the capability to protect devices accessing corporate data in both online and offline mode as follows:

- An **online device** is one that is actively connected to a network. This can be any type of network capable of transmitting data either to or from the device. The most common data network interfaces are Wi-Fi and standard mobile data networks (3G and 4G/LTE), though there are other ways of transmitting and receiving data on a mobile device, as shown in Figure 1-3. These include Bluetooth; short message service (SMS); multimedia message service (MMS); and tethering or synchronizing a device to another device, such as a laptop. When a device is online, your security deployment needs to protect data and applications on the device, as well as provide protection for data as it transits the network.

Over the next few years, many mobile operators will be transitioning from their current 3G networks to faster, higher-capacity 4G/LTE networks. Technically, Long Term Evolution (LTE) networks do not qualify as 4G, or Fourth Generation, networks, but many carriers market their LTE networks as 4G. In either case, these networks are significantly faster than the 3G networks they are replacing, opening up a huge wave of additional smartphone capabilities and, more than likely, additional security concerns along with those capabilities.

When online, you must protect the device regardless of the type of data connectivity it has.

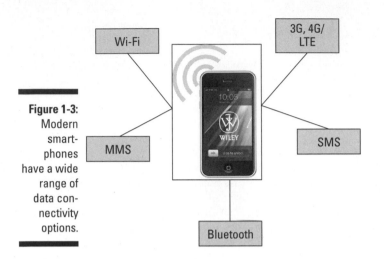

Figure 1-3:
Modern
smart-
phones
have a wide
range of
data con-
nectivity
options.

✔ An **offline device** is one that is not actively connected to any network. In this case, the potential *attack vectors* (methods by which a device can be accessed for malicious purposes) are limited because there is no way to get data onto or off of the device. Still, it is important to protect data and applications on the device. Loss, theft, and dormant malware are still issues to be concerned about with a disconnected device.

The techniques and technologies described in this book are targeted toward building a complete mobile device security strategy, one that will allow your organization to protect the data, applications, and devices themselves regardless of the type of network (or lack thereof) the device is connected to.

Applications Galore: Exploring Mobile Device Applications

Other variables that will impact your mobile device security strategy are the applications and data running on these devices. We define four types of applications (e-mail and messaging, web-based, client/server, and standalone) for the mobile device use case. Each type of application comes with its own set of security concerns, such as the ability to control who gets access to the application *(access policies),* as well as the ability to restrict specifically what each individual user may access within each application *(granular control)*, and all are addressed in this book.

E-mail and messaging

E-mail and messaging applications are among the most popular enterprise applications leveraged on mobile devices. The most common include e-mail send/receive, calendar, contact, and task synchronization. These applications are typically accessed via a Microsoft Exchange e-mail server (or similar).

Other messaging applications include chat or instant message, short message service (SMS), multimedia message service (MMS), and, potentially, video-conferencing applications.

The primary concern that enterprises have when enabling e-mail access from mobile devices is the loss or theft of the e-mail data. Enterprise e-mail can contain all types of sensitive information, from financial results to product designs. Sending that data to a mobile device that can easily be lost or stolen can be a scary proposition. We explore how to mitigate these concerns throughout this book.

Web-based applications

Every smartphone on the market today includes a web browser for viewing web pages and for leveraging web-based applications. In some cases, the application developer has optimized special versions of the application for mobile device access; in other cases, the web content is the same whether it is viewed on a smartphone or on a desktop PC. Regardless, these applications are unique in that they are accessed exclusively through a web browser, with no installed device application or other client-side component.

Despite the fact that web-based applications are hosted on a server in the network, there are still exposure and security concerns that you need to be concerned with, including the following:

- Some data might be downloaded and stored on the device.
- There is the possibility of man-in-the-middle or other types of attacks that can hijack or intercept the web application session and leverage that to steal data or to download malicious code to the mobile device.

Client/server applications

Client/server applications are traditional *fat client* applications, which require that the device has natively installed software to run the application. These installed applications communicate with application servers running inside the corporate network.

Until recently, there were very few client/server applications in use in a typical enterprise environment. Over the last few years, however, their use in the enterprise has really started to gain in popularity. As enterprises have embraced smartphones and tablets as productivity tools, and increasingly as primary devices, the need to allow users to access everything that they are able to access on their laptops and desktops has become prominent. As with other applications, these types of applications aren't without their security issues, so when rolling these out, ensure that your security strategy can protect the data associated with these applications.

Standalone applications

Standalone applications are those that function on the device itself, with no server-side or backend component. There are many such applications. In the enterprise, the most common applications in this category are office or productivity applications. Many of these applications have a web-enabled component, but they are primarily used for viewing and editing spreadsheets, documents, PDFs, and presentations.

The issue is that these are the types of files that typically contain your most sensitive corporate data. The security techniques described in this book will help you to ensure that you are properly securing this data, both when it is stored on the device and when it is transmitted to or from the device.

Allowing Smartphones onto Your Network

This book shows you that you need to take many factors into account when planning your mobile device security deployment. Because this solution spans multiple types of technology, you need to properly plan every piece of the process and follow those plans in that order.

In the following sections, we give you an introduction to the different components of a successful deployment. We discuss most of these topics in great detail throughout the various chapters of this book. Be sure to mark down areas that you and your organization might find difficult, and then find out more about those topics, either by referring to the appropriate chapters of this book or by doing your own organizational and self-directed research.

Educating yourself on the risks

Reading this book is a great start to educating yourself on the risks of allowing mobile devices on your network. Throughout this book, we go into detail about the types of risks facing mobile devices and provide insight into the ways that you can mitigate those risks in the real world. Chapter 14 provides a number of additional online resources you can consult to dig deeper in various areas. Additionally, the threat landscape changes often and quickly, so stay on top of the latest mobile security news.

Scoping your deployment

You may work in an organization where a large percentage of the employees need to access corporate data from their mobile devices. On the other hand, you might work in an organization where only a small portion of employees would benefit from the increased productivity that mobile device data and application access can provide. Regardless, limit access to only what's necessary; after all, every additional person who has access to corporate data on a mobile device represents another potential area for data to be lost or stolen. Prior to rolling out a solution to your end users, determine who should have access and from what types of devices, for instance, corporate- or employee-owned, and any operating system or only certain operating systems? This helps you contain the size of your deployment and limit access to devices and users where you feel you have a good handle on risk.

Creating a mobile device security policy

Your mobile device security implementation is only one piece in a broader corporate security policy that governs the technologies that are implemented to ensure proper security in your organization's network. This policy provides guidelines that you can follow when planning to allow mobile devices into your network; it also is a great reference for the implementation team as it evaluates vendor solutions and begins the deployment. Chapter 4 introduces the topic of mobile device security policies and includes an example from our ongoing AcmeGizmo case study. (See the next section for an introduction to this case study.)

Determining device configuration policies

Your mobile device security policy has an immediate impact on the types of configuration policies that you will apply to the mobile devices in your network. For example, the security policy might state that all devices must have a lock password with certain requirements. Or the policy might state that all devices must have full disk encryption. Either way, you need to decide on a detailed set of configuration policies while rolling out your mobile device security solution. We cover this topic in detail in Chapter 4 and show how our case study organization, AcmeGizmo, configures policies on its devices in accordance with its enterprise mobile device security policy.

Figuring out how you'll connect devices to your network (s)

Another integral part of your overall mobile device security strategy is connectivity to the corporate network. Your organization has most likely already deployed a VPN of some sort (such as IPSec VPN or SSL VPN) for remote access into the network from laptops running Windows, and perhaps even Macintosh and Linux. As you expand to mobile devices, you'll find that some VPN solutions support the wide range of mobile operating systems, and others do not. Therefore, you need to evaluate the current state of affairs and determine whether your existing VPN meets your needs as you expand the scope of your remote access solution. In addition to choosing a VPN, you need to make other decisions, such as what type of authentication to enforce from mobile devices and how much access to provide to users in various employee groups. These are critical decisions, and we cover this topic in detail in Chapter 7.

Devising an endpoint security strategy

The number and types of threats facing mobile devices are growing quickly as these types of devices become more popular and begin to contain much more sensitive, and potentially valuable, information. That is exactly why it's so important to deploy an endpoint security solution as you start to allow mobile devices into the network, just as you have probably done for traditional systems, such as laptops, desktops, and netbooks on your corporate networks. Antivirus and personal firewall capabilities need to be at the heart of your endpoint security strategy for mobile devices. These and other endpoint security technologies are discussed in Chapter 10.

Planning a strategy to deal with loss and theft

No matter how many policies you apply and how much security you enable on the mobile devices in your network, some of them will be lost or stolen. When such situations arise, you not only need technology to help you deal with these events but also require processes and procedures to deal with them quickly and effectively. Whether you allow users to track, lock, and wipe their own devices or whether your helpdesk team does that, everyone involved needs to know exactly what to do when a device is lost or stolen. Chapter 11 deals with this topic in detail.

Seeking vendor info and requests for proposals

After you educate yourself and identify your deployment team, look at various vendors to come up with a short list of mobile security vendors that you can invite in for further evaluation. Note that different vendors cover different areas of functionality, with no single vendor covering all possible functionality. You will likely need to deploy more than one product to accomplish all your organization's goals.

Different organizations take different approaches to narrowing the list of vendors. Some organizations initiate design/sales meetings with interesting vendors to see how each vendor implementation fits with their organization's goals. Other organizations create requests for proposals (RFPs) that give vendors a list of questions that they must respond to in writing. Regardless of the approach, the goal is to identify which vendors offer products that have sufficient functionality to meet the key goals of your mobile security deployment.

Implementing a pilot

You can gain a lot of information from deploying your mobile security solution to a small group of users prior to a wide rollout. Many plans have failed in the implementation phase due to unforeseeable issues. Start with a pilot group that consists of a small, but representative subset of your user population. Be sure that you include users with a variety of device types that you will allow into your network. Also, include members of every business group that will access the network because different applications and security requirements might result in different user experiences. When you add end users to the equation, you get a good sense of how seamless the mobile security solution will be for users as a whole, how well your organization can deal with deployment problems, and whether the chosen vendor can meet your needs.

Assessing and reevaluating at regular intervals

Congratulations! You've successfully rolled out your mobile device security implementation, and your users are all happily connecting to the corporate network from their devices of choice — iPhones, Android devices, and so on. So, now what? Vacation? Retirement? Put your feet up on the desk? Not so fast; network, security, and user requirements evolve over time. While these changes happen, your mobile security strategy must also change. After you complete the deployment, ensure that your users are happy, that your team can effectively manage and support the deployment at scale, and that as threats to mobile devices evolve over time, your mobile security solution continues to meet your organization's security goals. Continual reassessment is a key part of any technology adoption, and you need to make it a part of something as critical and visible as mobile devices.

Introduction: AcmeGizmo Enterprise Smartphone Deployment Case Study

Many of the chapters in this book end with a case study. The storyline is ongoing and follows a fictional company named AcmeGizmo. At this organization, much like at many other organizations, employees have widely adopted mobile devices. You join the story as Steve, the CIO, asks Ivan, the IT manager, to come up with a strategy for securely allowing these devices into the network.

Every chapter that includes a case study takes a close look at the decisions that Ivan makes in order to accomplish this goal.

AcmeGizmo is a vertically integrated, global manufacturer of widgets. Its 8,000 employees span the range from employees on the manufacturing line in their factories to a retail sales force working in their stores and kiosks.

Exploring legacy smartphone deployment

Historically, AcmeGizmo has provided many executives and salespeople with company-issued BlackBerry devices. From these devices, employees can access their e-mail, calendar, and contacts, in addition to a few select intranet sites. Here's an overview of the network:

✓ **BlackBerry Enterprise Server (BES):** AcmeGizmo has been very comfortable with the secure nature of its BlackBerry deployment. In addition to the devices themselves, it has a BlackBerry Enterprise Server (BES) in its network that helps it manage and secure these devices. The BES provides VPN and authentication for the BlackBerry devices, which securely connects the remote device to the network and secures all data as it transits the network to the corporate data center. In addition, the BES is the primary tool that AcmeGizmo's IT staff uses to manage policies and configure BlackBerry devices. The policies range from password complexity policies to application provisioning and blacklisting policies.

✔ **Connect PC VPN:** AcmeGizmo has a fairly standard deployment for remote laptops, all of which are Windows-based. For this, it has an IPSec VPN appliance from Connect PC. The VPN appliance handles encryption and authentication, in addition to several other critical remote access features. AcmeGizmo also has a suite of desktop management tools that help it manage policies on the remote laptops and ensure that those machines are appropriately patched and configured. (Connect PC is a fictional company.)

✔ **Secure PC Endpoint Security Suite:** AcmeGizmo has invested heavily in an endpoint protection suite from Secure PC, which includes antivirus and personal firewall software. (Secure PC is a fictional company.)

The only devices permitted to access the AcmeGizmo network remotely are AcmeGizmo-owned and -managed BlackBerry devices and Windows laptops. Figure 1-4 shows the legacy AcmeGizmo network, with BlackBerry devices connecting to the BES, and Windows laptops connecting through the Connect PC VPN. A *legacy network* is the network that AcmeGizmo put into place prior to implementing the mobile security strategy that Ivan designed as a result of reading this book.

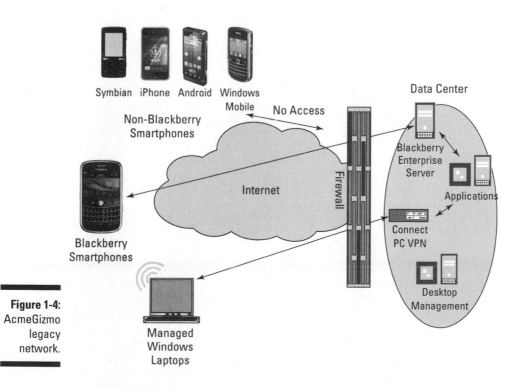

Figure 1-4:
AcmeGizmo
legacy
network.

Enter the smartphone explosion

About a year and a half ago, however, a rapid shift began to occur. It all started the day that Brooke, the CEO, purchased the latest and greatest iPhone. Upon arriving in the office the next day, Brooke stopped by Steve's office and essentially demanded access to her e-mail from the new iPhone. Not wanting to upset his boss, Steve asked Ivan to figure out how to make this happen as quickly as possible. The challenge was that a lot of AcmeGizmo's IT security investments were of little use when attempting to secure the device and its access to the network.

For its desktops and laptops, AcmeGizmo has traditionally worked with Secure PC, the leading endpoint security vendor, to purchase its entire suite of functionality, including antivirus and personal firewall components. Unfortunately, Secure PC's sales rep confirmed for Ivan that it doesn't currently have a solution for smartphones, but that it's "on the roadmap," offering very little help for Ivan in his current situation.

Ivan next took a look at AcmeGizmo's VPN platform from Connect PC, the leading IPSec VPN solution. As with Secure PC, Ivan quickly determined that Connect PC had not added support for smartphone platforms like the iPhone.

As Ivan's frustrations began to mount, the number of issues that he anticipated began to grow. He realized that none of his systems could properly handle this smartphone problem. He had no way to enforce appropriate configurations on these devices, as the BlackBerry Enterprise Server was extremely feature-rich but covered only BlackBerry devices. He had no way of controlling or enabling application distribution, no way of wiping data from a device if it was lost or stolen, and so on.

So, Ivan took a step that many IT managers have taken. Under immense pressure from his boss and from the CEO, he deployed the mail server so that it was directly accessible from the Internet, enabling the CEO to access her e-mail, calendar, and contacts. Luckily, the mail server included some functionality to control the device itself, including the ability to set a password requirement and remove sensitive data from the device if it was lost or stolen. Unfortunately, that was about it. Many of the advanced policies and the layered security approach that Ivan had spent a lot of money and countless hours deploying for AcmeGizmo's laptops and desktops were useless for these smartphone platforms. Ivan felt that what he had deployed was somehow insecure, and that he was taking a great risk by allowing the CEO to access so much sensitive data from her iPhone. Figure 1-5 shows the revised network, which allows the CEO's smartphone to access the network.

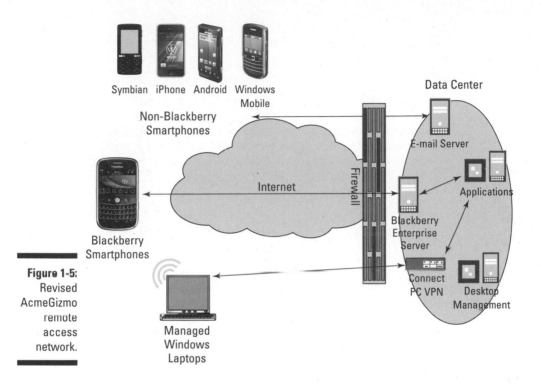

Symbian iPhone Android Windows
                                Mobile

Non-Blackberry
Smartphones

Data Center

E-mail Server

Internet   Firewall

Applications

Blackberry
Smartphones

Blackberry
Enterprise
Server

Connect
PC VPN

Desktop
Management

Figure 1-5:
Revised
AcmeGizmo
remote
access
network.

Managed
Windows
Laptops

As the next couple of months went by, Ivan began to notice an increasing number of unapproved mobile devices in the company cafeteria, in meetings, and in the hallway. This made him suspicious, so he went back to take a look at the mail server logs. What he found surprised him — there were more than 500 unapproved devices accessing the mail server without his knowledge! Word had gotten around that this type of access was possible, and people had started using it.

Upon reporting this finding to Steve, the CIO, Ivan was asked to drop everything and make mobile device security his top priority. It was apparent that end users wouldn't easily give up their new smartphones, and the increase in productivity was something the company would definitely benefit from. Plus, Steve viewed it as a cost-savings measure. Many of these users were previously on company-sponsored phone plans with their BlackBerry smartphones and were now paying for their own service. The company BlackBerry bill had gone down by almost 20% in the past three months!

Ivan was asked to figure out a mobile device security strategy for three different groups of employees: executives, enterprise salespeople, and, generically, all other employees who have mobile devices and wish to access e-mail and data from them. Throughout the book, we discuss the choices that AcmeGizmo has made in several key areas, corresponding to the following chapters:

✔ In Chapter 4, we illustrate the security policies that Ivan has developed for the smartphones accessing the AcmeGizmo corporate network.

✔ Chapter 5 ends with a discussion on AcmeGizmo's control of applications, as well as its struggles with whether to monitor employee use of smartphones.

✔ Chapter 6 discusses some of the special concerns that arise when AcmeGizmo attemps to provide access to some members of the Finance department who have access to point-of-sale transaction data for their retail stores. Ivan has a particular concern about their ability to meet Payment Card Industry (PCI) Security Standards Council compliance standards.

✔ In Chapter 7, AcmeGizmo decides to move forward with an SSL VPN solution, a consolidated product that can provide remote access from Windows laptops and various smartphones at the same time, simplifying operations.

✔ The case study section in Chapter 10 shows how AcmeGizmo leverages a mobile security offering to embed endpoint security into every one of the smartphones accessing its corporate networks.

✔ Chapter 11 illustrates what happens the day that Ed in engineering loses his smartphone, which has the company's next-generation widget designs on it (certainly something AcmeGizmo does not want getting into the hands of its competitors).

✔ Finally, in Chapter 12, we talk about the backup and restore strategy that AcmeGizmo implements for its smartphone devices.

Chapter 2

Why Do I Care? The Mobile Device Threat

*I*n the present day, employees are king, bringing with them (into the network) not one, not two, but sometimes three or more personal devices that have little or no corporate-approved applications; and yet they connect to the corporate network and chat, e-mail, talk, network socially, and connect to the cloud. It's a bad horror movie, *The Invasion of the Devices,* and you're the hero who's being overrun.

These devices are highly customizable (unlike enterprise-issued laptops, which typically have a lot of restrictions tied to them in terms of what applications the employee can install). Therefore, the employee has a personal attachment to these devices and swears by them both at work and outside of work. And this phenomenon, which could be brushed off as an anomaly just a few years ago, is fast becoming the norm in enterprises, so you need to take notice.

Mobile device security isn't a problem that you can just wish away. Employees will do things they shouldn't, such as pick up malware from a free app they just downloaded to their BlackBerry. That leaves you — the IT professional with corporate responsibilities — to be accountable for preventing security breaches where possible and remedying a breach after it happens.

This chapter helps you understand the threat posed by unsecured mobile devices and explores the tools available to help secure them. In this chapter, you discover what the impact of smart devices in the enterprise means for maintaining data integrity, network utilization, user productivity, secure communication, device manageability, and compliance capabilities. And you find out how taking an assessment of security challenges is compounded by the device invasion. And finally, you discover what measures you need to put in place to secure the new enterprise mobile environment. To do that, first you must understand the problem that exists and make a case of why you should not quit, but form a plan to assimilate the invading hordes.

Recognizing the Scope of the Threat

If you have been on any planet but Earth, you can be forgiven for not having noticed the smartphone explosion. The rest of us who exist in Earth's modern, connected society recognize this phenomenon. Smartphones will soon be arriving at your workplace in droves, if they aren't already there.

A smartphone is just one type of mobile device that may show up in the workplace. Employees may also use other mobile devices (netbooks, tablet computers, or any other form of Internet-connected device) on a daily basis.

Despite the influx of mobile devices, their mere presence in the enterprise is not the problem. But considering the habits and practices of mobile device users who co-mingle work and personal activities helps you begin to outline the scope of the problem. For example, the devices your company's employees use to read their work-related e-mail may also be the devices they use to post pictures and status updates on Facebook. Such practices expand the scope of your company's responsibility for managing and monitoring mobile device use.

Consider the following interconnections and interactions that happen through mobile device use.

Loss, theft, and replacement

Your employees' mobile devices may change hands for a number of reasons, exposing your company data to others.

Three main characteristics define the demise of a mobile device:

- ✔ **Loss:** Mobile devices are tiny, and your employees can lose them a lot easier than they can a desktop computer. Mobile devices can easily slide out of your employees' pockets or purses.

- ✔ **Theft:** These devices are very attractive to thieves because of their popularity and resale value.

- ✔ **Replacement:** Your employees like to periodically upgrade their old phones to newer, sexier devices, and, as a result, sell or give away the older devices. So why do you care? Because these devices frequently contain proprietary enterprise information that can fall into the wrong hands.

Lost or stolen devices are ticking time bombs until they can be deactivated. Unscrupulous folks who have possession of these devices can access your network and assets inside your network. So the exposure is very high.

Most device manufacturers and OS vendors as well as some third-party software vendors offer what is commonly referred to as mobile device management (MDM) capabilities (described in detail in Chapter 5). In the context of lost or stolen devices, this typically entails locating the lost device, wiping out sensitive data from its memory, and preventing it from attaching to the network.

Some tech-savvy users can take matters into their own hands using software like the iPhone's MobileMe capability to do the locate, wipe, and lock out actions; others can call the service provider who are also increasingly offering MDM service, too. You can educate the users about these avenues to prevent a lost device from becoming a liability to you.

It may be your employee's device, but it is your company's data that it is carrying.

Really off-site data storage

The exploding storage capabilities of mobile devices — which is further augmented by applications that extend storage to the cloud — present a growing possibility of intellectual property and sensitive information being widely downloaded and stored and, more critically, compromised.

The *phone* in *smartphone* belies the capabilities of these devices. Your employees frequently download all kinds of enterprise data (spreadsheets, presentations, e-mails, and so on) that are stored in these phones with ever-expanding *memory footprints* (the amount of memory used by applications). Such use makes these phones an IT asset that needs to be guarded as zealously as servers or other storage devices.

Figure 2-1 shows the amount of memory available in various smartphone models. Multiply these numbers by the number of suspected smartphones in your enterprise, and you can quickly see how much company data could be downloaded onto this mobile storage. It's enough to give you (as someone tasked to secure data) the heebie-jeebies.

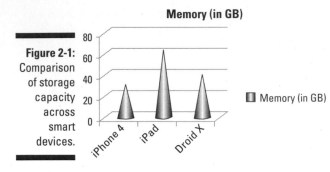

Figure 2-1:
Comparison
of storage
capacity
across
smart
devices.

Free (but not necessarily nice) apps

With the advent of free and nearly free applications available for download for every smartphone, your employees are experimenting with new apps all the time. After all, don't you?

Almost exclusively, these applications are designed with the consumer in mind. So the fact that cheap and free apps lure your employees into constantly experimenting and downloading these apps to their devices isn't surprising. And such experimentation results in devices that are constantly morphing and being exposed to potential malware. This situation is quite different from other enterprise devices such as the venerable desktop, or even the laptop, which has a more stable countenance, and therefore affords you a baseline to protect against.

It would behoove you to establish an approved set of application types and versions that could be your baseline for mobile devices. This allows you to evaluate any deviations from this baseline as your users customize their devices.

This is a rapidly changing landscape, and you need to keep abreast with the latest application sets all the time; otherwise, your users will be either very unhappy (because you disallow new apps that are not part of your baseline) or they will be angry (because you're seemingly ignorant of the newer apps). Either way, you are in serious danger of renegade applications running amok.

Network access outside of your control

By their nature, mobile devices connect wirelessly to available networks, most of which are outside your company's control. The proliferation of wireless interfaces means an ever-increasing attack surface that can be used to compromise mobile devices.

Smartphones have a variety of radio interfaces to cater to the always-on connectivity that users demand, as well as the need to use the smartphone as a hub to connect to accessories, such as a stereo headset, GPS Bluetooth receiver, Wi-Fi hotspot, and so on. While these interfaces enhance the user's experience, they also expose your company to yet another *attack vector* that the bad guys are waiting to exploit. (An attack vector is a mechanism that is used by the attacker to gain access to a critical resource in order to deliver malware or compromise the entity.) Radio interfaces are a problem, not a feature. So again you're faced with the dilemma of how to rein in the runaway vendor manufacturers cramming in more and more radio interfaces that your users are actively lapping up and exposing their smartphones to risk on a continual basis. Figure 2-2 shows the multitude of radio interfaces on today's mobile devices, each one further exposing the device to potential attacks.

Figure 2-2:
Some of the radio interfaces on the smartphones today.

It is only a matter of time before mobile devices become *multihomed,* meaning they are connected to multiple wireless interfaces simultaneously. Therefore, you need to be aware of and protect against all of these interfaces simultaneously. Chapter 10 goes into detail about what tools you can use to protect devices against these attacks.

Understanding the Risks

Now that you've explored the characteristics of a mobile device and its implication on security, it is time to delve deeper into the risks you face from a compromised mobile device.

Opening the door to hackers

First, we examine in some detail how these mobile devices can get compromised to begin with.

The get out of jail card

Given that the primary focus (so far) has been the younger generation, who tend to spend more on mobile gadgets and renew devices faster, it isn't surprising that ease of use trumps everything else. Nowhere is this more evident than the website that goes by the illustrative name of JailbreakMe 2.0 (www. jailbreakme.com).

Jailbreaking is the term used for hacking into one's device and freeing it from the controls imposed by the device manufacturer. While jailbreaking phones is nothing new, the website JailbreakMe 2.0 is interesting because it succeeds in opening up the phone with the user simply visiting the website, as shown in Figure 2-3. There are no special loaders, no rebooting the phone into recovery mode, and no connecting the device to a computer. Users just visit the site on their iPhone, iPad, or iPod touch, confirm that they want to jailbreak the device, and then sit back and wait. This is convenient for users who want to jailbreak their phone.

Figure 2-3:
Jailbreak with an easy slide.

Note that jailbreaking is different from *unlocking* a phone, which essentially frees it from the carrier's controls and allows a user to connect to another carrier's network. Jailbreaking, on the other hand, frees the device from the controls of the device manufacturer and allows the user to download unvetted (by Apple) applications.

Unfortunately, the ease of which a user can jailbreak a phone through a site like JailBreakMe is also very convenient for an attacker. The ability to cause code to be executed on a device with high privilege simply by visiting a website is the very essence of a drive-by attack. The JailbreakMe website kindly asks the user to confirm that his device should be broken, but an attacker is unlikely to be so gracious. The fact that these devices can be opened up so easily represents a serious security flaw.

Some reasons for jailbreaking include a user wanting to choose what applications can be installed (and uninstalled), which carrier's SIMs the phone can support, and what administrative settings can be imposed on the device. With the recent ruling by the U.S. Library of Congress that jailbreaking is legal (July 2010), the legion of users who will experiment with jailbreaking will only increase.

The Digital Millennium Copyright Act is a U.S. copyright law that criminalizes the production and dissemination of technology and services that circumvent measures protecting copyrighted works. However, the U.S. Library of Congress decided to add exemptions to this act, specifically "that Americans are within their rights to modify, remove or replace software on their mobile devices." What this means is that your users are well within their legal rights to jailbreak.

So what went wrong? How did the perpetrators break through the impregnable walls of the much lauded iOS and iPhone? Well, it seems that the attackers are exploiting at least two security vulnerabilities:

- **Heap overflow bug:** The attack makes use of a classic heap overflow bug in the Type1 font decoding of Apple's PDF viewer. The attack payload consists of a PDF containing a malformed font that exploits the bug and executes code in the Safari browser's memory space. Once code is running in the browser's space, it then has to break out of the sandbox.

  A *sandbox* in this context refers to an isolated application environment that prevents one application from encroaching on another application's environment. It is a commonly used technique to provide compartmentalization leading to better security.

- **IOSurface library bug:** As far as we can tell, the attacker breaks out of the sandbox by exploiting another bug, this time in the IOSurface library that handles various types of video and OpenGL rendering onto the screen. A set of parameters is passed to IOSurface with values that look suspiciously like ARM machine code instructions, along with various values that look like they're designed to trigger heap overflows. Because the IOSurface code has direct access to hardware, it has to run outside the sandbox. Therefore, by triggering the execution of attack code from within this library, the attacker can break out of the sandbox and control the whole phone.

If a highly regulated entity like Apple, which thoroughly vets both its in-house products and third-party products, can be the subject to this kind of subversive attack, imagine what a more "open" operating system like Android can be subject to. By the time you read this, there are likely to be many breaches and patches.

App (did we say malicious?) Store

At the highly regarded Black Hat DC 2010 conference (a premier security event about security breaches and vulnerabilities in hardware and software), the principals of the App Genome project (which focuses on the Android platform) revealed some disturbing facts. They found a series of wallpaper applications that gathered seemingly unnecessary data — a device's phone number, International Mobile Subscriber Identifier (IMSI), and the currently entered voice mail on the phone number. And here is the kicker: The apps were transmitting this data unencrypted to a server.

Would you feel compromised if your employees' information could not only be accessed by a seemingly innocuous application but also shared with some untrusted third party?

Another recent development in app security breaches was on the App Store for iPhone. A rogue application developer figured out a way of hacking into other members' iTunes accounts and initiating fraudulent purchases of all of his apps. At one point, his applications accounted for 42 of the top 50 apps on App Store.

Writing malware for a mobile phone

Lest you be lulled into believing that it takes a PhD to write malware for the mobile phone, let us blow that fallacy away. An experiment conducted by BBC News along with security firm Veracode involved creating a crude game for a smartphone that also spied on the handset. It wasn't hard to do. In fact, the application was built using standard parts from the software toolkits that developers use to create programs for handsets, which also makes these malicious programs hard to identify because they use the same building blocks as benign programs.

What's worse, creating the program took only a couple of weeks. The program included a crude game that surreptitiously gathered contacts, copied text messages, logged the phone's location, and sent all the information to a specially set up e-mail address — all built from standard library functions that legitimate programs use.

Compromising your business communications

Next we move on to security issues that prevail upon the very lifeline of enterprises: communication (voice, data, text, instant messaging, and so on).

Electronic eavesdropping

Societal decorum forces us to talk in suppressed tones while using our smartphones and devices in public places, often because we don't like to be eavesdropped upon.

But the same cannot be said of electronic eavesdropping, which is easily achieved with the tools available today. It's the easiest way for your employees' devices to be compromised. When users inadvertently download an infected application, malicious spy software can be surreptitiously installed to collect data and voice, and forward it to a rogue server that can then analyze the sensitivity of the information and decide how best to use this proprietary data.

The regulatory environment that you are in can vary based on geography and industry, so be aware of the impact of trying to secure the communication channels to prevent eavesdropping from happening, as well as the fallout of the result of the communication channel having been breached by a malicious third party.

SMS (also known as texting) — it can be hacked

The number one application for users and mobile operators worldwide is SMS (short message service) — that is, text messaging. The popularity of this simple yet effective service is really remarkable, and Figure 2-4 indicates that the top application among U.S. adults really is SMS.

Couple the popularity of SMS with the explosive growth in the U.S. (see Figure 2-5), and it becomes apparent that SMS is a great magnet for someone with mal-intent wanting to take advantage by eavesdropping.

Although traditional SMS uses GSM encryption that has been cracked only in university research labs and is relatively secure, there are increasing indications that this veneer of security around SMS is starting to crack. (See Chapter 10 for more on GSM encryption.) The ability to turn a phone into a surveillance gadget to capture text messages is possible. One way to do this is *phone cloning*, which allows somebody to impersonate others' phones and masquerade as the original. This obviously works with only one phone at a time, but you can imagine your fate if that one phone happens to be the CEO's, or yours for that matter. For the hackers who are interested in casting a wider net, there are ways to load illegal firmware onto one's device so that it can listen in on multiple radio channels to pick up text messages that are sent from other devices that are using the same channels.

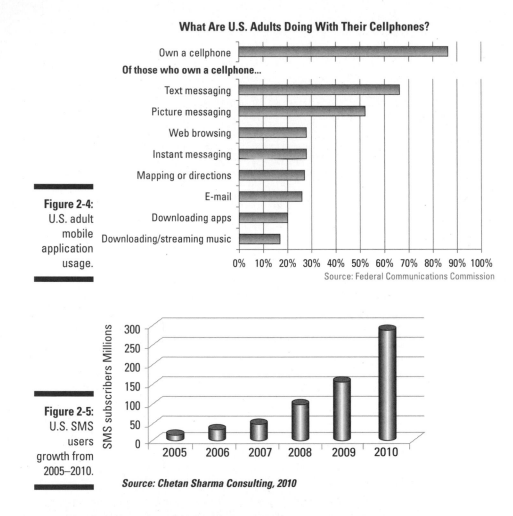

What Are U.S. Adults Doing With Their Cellphones?

Figure 2-4:
U.S. adult mobile application usage.

Source: Federal Communications Commission

Figure 2-5:
U.S. SMS users growth from 2005–2010.

Source: Chetan Sharma Consulting, 2010

Endangering corporate data

Shifting gears, we now focus on where your intellectual property may actually be doing the rounds. Given that users have access to corporate data across a variety of mobile devices, keeping tabs on where this data resides at all times is crucial. And any vulnerability or attack on the mobile devices inherently endangers the corporate data. The following sections look at the different ways that this data could be compromised.

Mobile data on the move

If you compare the specifications of the four top smartphones, as detailed in Table 2-1, the statistic that jumps out is the fact that they compete along the matrices of processor speed, size of display, camera resolution, memory size, and creating a roaming hotspot.

This a big problem for you. Imagine having your employees walking around with hotspots in their pockets, carrying sensitive enterprise data. The *Invasion of the Devices* is now faster, bigger, fatter, and better.

Table 2-1	Comparison of Some Popular Smartphones			
Feature	*Motorola Droid X*	*HTC EVO 4G*	*iPhone 4*	*Samsung Vibrant*
Camera	8MP	8MP	5MP	5MP
Processor	1GHz	1GHz	1GHz	1GHz
Display	4.3 inches	4.3 inches	3.5 inches	4 inches
Storage	8GB internal	8GB internal	16GB/32GB internal	16GB internal
OS	Android 2.1	Android 2.1	iOS 4	Android 2.1
RAM	512Mb	512Mb	512Mb	512Mb
Wi-Fi hotspot	Up to 5 devices	Up to 8 devices	Up to 5 devices	N/A

The highly consumer-focused view of the world results in devices that are flashy, are user-friendly, sport gorgeous displays, and have large amounts of memory but very little in terms of features that you as an IT professional might care about, such as hardware encryption, built-in locators, remote wipeout, and lockdown capabilities.

Server resident data

One trend that has significant ramifications for your employees' data is the storage of content on servers operated by content and service providers. The applications that your employees download are increasingly augmenting the storage that is available on the device by using a server extension. It's a convenient way to provide a backup for the device itself. But imagine the scenario in which your employees are downloading sensitive corporate data to their devices, and this information is then periodically synchronized with the servers in the service provider's domain. This situation not only violates all of your typical enterprise security policies but also lends itself to potential compromise of the data in the cloud that could come back to bite you. Figure 2-6 shows four popular cloud storage apps for the iPhone, highlighting the ease of use of these cloud storage apps, which means nothing but migraines for you.

Cloud, in this context, refers to cloud storage, which is an online-hosted storage service provided by third parties to augment or replace any local storage that you may host.

Figure 2-6:
iPhone cloud storage apps.

There was a well-publicized incident involving the T-Mobile account of a celebrity's Sidekick device wherein all of this person's address book data, photos, e-mail, and voice mail maintained in the cloud was compromised. Unauthorized users gained access to this information and had a field day propagating it. While this event was limited to one celebrity's personal trauma, imagine if this were a high-profile employee of your company and the same breach were to occur.

Reckless user experimentation

Jailbreaking, as described earlier in this chapter, is the term used for hacking into a device and freeing it from the controls imposed by the device manufacturer.

While jailbreaking certainly is welcome news to freedom-loving users who are glad to be rid of the shackles imposed by their device vendors and carriers alike, make note that it brings forth a brand-new set of headaches for you. For instance, the typical jailbroken iPhone has the SSH server running by default, and worse still, has the default password of *alpine.* Now it doesn't take a genius to launch a probe at a hotspot or other commercial establishment for jailbroken iPhones and ssh into the device and wreak havoc. In fact, one of the earliest exploits of this was a Dutch hacker who did exactly that and left the hapless user with a screen similar to Figure 2-7.

Figure 2-7:
A ransom note posted on jailbroken iPhones.

Going to the website directs the user to send five euros to a PayPal account, after which the hacker will e-mail instructions to remove the hack, which most likely involves restoring the iPhone to factory settings. Now in relative terms, this is not a serious attack, and any compromised iPhone user will more than likely be willing to pony up five euros to get on with his life. However, the underlying vulnerability that is exposed by the user experimentation is the worrying part.

Another variant of this type of attack involved a hacker taking over the background screen of the iPhone and posting an image of yesteryear's pop singer Rick Astley, as shown in Figure 2-8. Annoying for sure. Imagine *Invasion by Alien Rick Astleys*.

While the risk of a hacked phone is quite obvious, what isn't as obvious is how a compromised phone could play havoc with enterprise data. For instance, all the data stored on the phone is now at the mercy of the malware. Similarly, any enterprise applications that can access corporate servers are also now vulnerable. One of more common pieces of malware turns on voice recording when the user is on the phone. Imagine a conversation between your employee and the CFO on the revenue numbers for the current quarter being surreptitiously recorded and spied on by unscrupulous third parties. The party is just starting, so get on your dancing boots.

Figure 2-8:
Rick Astley
in the back-
ground of
a hacked
iPhone.

Secure Shell (SSH) is a network protocol that allows data to be exchanged using a secure channel between two networked devices. It's a common adminstration tool that is used to configure and troubleshoot remote devices. In an infected device, the recommendation is to navigate to /private/var/ mobile/home, which hosts the viral files. Files named inst, cydia.tgz, duh, sshd, and syslog should be removed to deactivate the malware. As you will see later in this chapter, the manual process of removing these files on hundreds of thousands of enterprise devices is challenging, and there needs to be a more automated and scalable way of doing this. Fortunately, there is one. Keep reading.

Infesting enterprise systems by using location-based services

LBS (location-based services) has been widely recognized as the one of a handful of types of mobile applications that have the potential to become the next SMS in terms of user-base penetration, revenue opportunity, and customer adoption. LBS is any type of service (typically delivered on a mobile device) that utilizes the location of the device to customize the service offering. Examples include coupons from nearby restaurants, instant location-based social networking, and so on.

Why is this technology so hot? That's simple — the ability to track the location of users and serve them with customized information that is of value at that time and place is an unbeatable proposition. SMS is arguably the most enduring and consistently high-revenue-generating service that providers globally continue to monetize. However SMS's ubiquity is not without its issues. SMS-based spam is one such irritant. Figure 2-9 shows examples of SMS spamming on an Airtel (an India-based mobile operator) customer's phone.

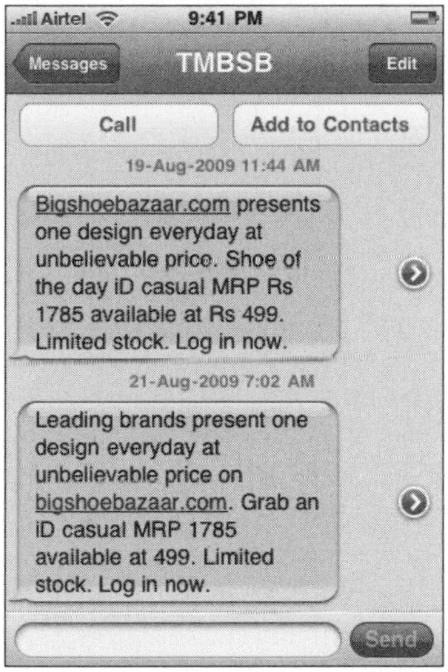

Figure 2-9: SMS spamming from an online retailer in India.

What does this mean for LBS? Can it be exploited the same way as SMS? You're reading this book, so you know that with every mobile opportunity comes the possibility for exploitation. Mobile users constantly trade privacy for convenience. The reason why LBS can be much more intrusive than SMS is the ability of the service provider to obtain the user's location. The convenience of a service that the user really wants may be a reason that he is willing to trade his location details. However, given the large number of applications that mobile users download regularly, individual location information that is used by each application is not always known. It is certain that some of these applications could not only use the location information for pushing unwanted ads or messages but also share this location information with third parties. The third party could then spam the user, and worse still, the spam could be a veiled phishing attack that compromises the device and its contents. That is why location-based spamming is so problematic. Say your employees are traveling around with smart devices that talk to the network all the time, advertising their coordinates. Spammers salivate for this kind of information so they can serve up customized spam to a targeted user base and veiled phishing attacks that could cause even the most wary and alert users to succumb.

Thwarting spam-only applications

There have been some positive developments on the vendors' front to thwart spam-only applications. A *spam-only application* is an unsolicited intrusion into the user's device by proffering up services — some legitimate (and solely out to make a quick buck) and others nefarious (trying to lure the user and compromising the device in the process). In fact, at Apple's iPhone development center website, it clearly states this:

> *If you build your application with features based on a user's location, make sure these features provide beneficial information. If your app uses location-based information primarily to enable mobile advertisers to deliver targeted ads based on a user's location, your app will be returned to you by the App Store Review Team for modification before it can be posted to the App Store.*

While it's encouraging to see this proactive stance by the regulator of the most popular App Store in the universe, there will always be more fertile ground for the spammers that is less likely to be as regulated, such as the Android Market, where spammers can sell their wares. And even in the Apple App Store, there is opportunity for spurious applications to circumvent the controls. Once Apple catches on, that loophole will be closed, but another will emerge and be exploited, and the cycle will go on.

At the face of it, location-based spamming does not seem like much of a problem because your users are being spammed all day long, anyway, and have tools and the common sense to filter these out. However, mobile spam is much more sinister in nature because the spammers use the location information to lure individuals, tempting them to hit an MMS or SMS or web page pop-up that seems very topical. For instance, your employee is driving by a movie theater on Friday evening and gets an offer for a coupon for a free drink and popcorn if she hits the accept button. When she does, her phone is infected by a Trojan that controls her smartphone and, by extension, your data, if she has connected to the network.

Assessing the Arsenal

In earlier sections of the chapter, we scour the mobile landscape evaluating the device characteristics, the infection vectors that could get a device compromised, and the risk of having compromised devices in your network. Now it's time to look at your arsenal of tools and find out how you can fight back against this growing threat.

To manage or not to manage

Say, for example, that until recently you have been issuing handsets to employees on behalf of the enterprise and, therefore, have been legally authorized to install monitoring, troubleshooting, and security applications at the behest of the corporation. Now, increasingly, your employees are demanding that they bring their own devices into the enterprise. This poses an interesting dilemma. Is an employee-owned phone accessing the enterprise network to connect to and use enterprise data considered an enterprise phone? If so, the legal definition of an enterprise-owned asset can be applied; and monitoring, security, and remote management capabilities can be handily used on the employee-owned handset.

If, on the other hand, this device does not fall under the purview of an enterprise asset, you have a problem on your hands because you have a bevy of uncontrolled and unmonitored devices roaming freely in your enterprise. Remember, the devices were bought by employees, and denying them access would be enough to cause a riot. However, expressly installing applications on each and every device is not only impractical, given the breadth of devices and the manual implications of such an endeavor, but also tough to sustain (given your shrinking IT budget).

So the takeaway here is that you need to be cognizant of this dichotomy and embrace a security policy that covers both enterprise-owned and employee-owned devices with a consistent view of protecting enterprise assets.

Chapter 4 goes into detail about how you can craft enterprise security policies that take into account these two classes of devices: enterprise-issued managed devices and employee-owned and -customized devices.

Where the need for compliance comes in

Do you work in a highly regulated environment, saddled with Sarbanes-Oxley, PCI, HIPAA, and the more recent bevy of financial regulations? Technically, your brethren in the utility, oil and gas, and airline industries are also saddled with regulations. So while you may feel burdened with staying in compliance, it is not a unique challenge to the IT industry.

Some of the weakest links in your enterprise asset chain are all those mobile devices. It puts your work processes in a precarious position because it leaves you vulnerable in the event of an audit, not to mention the larger business is at risk as well. The need to protect PII (personally identifiable information), confidential client information, and proprietary company data are the key tenets of many of these regulatory requirements, and the surge of mobile devices is eating through your compliance.

Mobile security apps start to emerge

Companies are starting to realize that you need to be armed with tools to protect your enterprise and its users. Companies such as SMobile (acquired by Juniper Networks in July 2010), Lookout, and others have security applications that are available on a variety of smartphones with the express purpose of providing a secure experience, including antivirus, URL filtering, malware detection, encryption capabilities, and so on. In addition, mobile-specific applications like SMS are also getting attention. Diversinet offers an SMS solution with robust encryption and authentication to augment the somewhat weak security that is built into SMS.

This means that the tools for enterprise mobile device security are increasingly available. You can start to experiment and roll out solutions to enterprise network users. The ubiquity of devices and their operating systems means that any solution that you employ needs to support at least the top smartphones in the enterprise so that you can cover the broad customer base with smartphones.

Planning to Sustainably Keep the Threat at Bay

The mobile device security problem is real and likely to get worse unless you take proactive measures. And because an after-the-fact security plan is much harder to implement, it is better to take the time now to define the devices, operating systems, and management applications that you will support and what levels of access you want to provide based on the *posture* (that is, the real-time assessment of the security grade of the device, taking into consideration the patch level and operating system version and the applications as well as any add-on features turned on by the user), location, and the device itself. In the following sections, we introduce some defensive postures that you should consider and then go into more detail in other chapters of the book.

Establish enforceable policies

However sticky this may be, the need to ensure that basic policies are enforceable on the devices accessing the enterprise is one of your fundamental challenges. Typically, these policies would be in the realm of

- ✔ Browser usage
- ✔ Video player usage

- ✔ Camera and video camera usage
- ✔ Installation of approved and unapproved applications
- ✔ Use of screen capture
- ✔ Use of location information
- ✔ SMS usage
- ✔ Connecting via Bluetooth
- ✔ Connecting via Wi-Fi
- ✔ Tethering
- ✔ Use of encryption
- ✔ Use of cloud backup services
- ✔ Use of local removable storage

Chapter 4 delves into the details of policy implementation.

Policy needs to be a constantly evolving template that you update with the rapidity of evolving devices, operating systems, and applications. Therefore, a "create it once and forget it" mentality is not a viable modus operandi. For example, you might need to add a jailbreaking policy to deal with a practice that may not have been in existence when you first created your smart-phone policy definition. With the landmark judgment passed by the Library of Congress in August 2010 that essentially makes jailbreaking legal in the United States, you now have to adapt your policy with the assumption that an increasing number of jailbroken phones will be winding their way into your enterprise. Don't be surprised by a breach or loophole that is exploited because you don't have a policy that covers this relatively new phenomenon.

Evaluate tools without biases

While it may seem very comfortable for you to go after security solutions that you have already deployed in the enterprise, from a vendor sourcing perspective realize that smartphones are very different beasts. Do yourself a favor by looking at the new mobile vendors who have offerings tailored for smartphones. Take into account the smaller memory footprint, battery life, location awareness, and myriad wireless interfaces, each presenting a unique attack surface. Also, the highly customizable nature of these devices means that any solution that you employ needs to be adaptable to suit the changing posture of the end device and to ensure that the device is secured at all times, or you should at least be able to flag deviations to policies so you can take remedial action.

Secure the location

Even though it sounds superfluous, it still bears repetition that the single most important attribute that a smartphone has — compared to its fixed counterpart — is its ability to move. Therefore, its location attribute is its crown jewel. This has not been lost upon vendors and app developers alike. Compromising a smartphone's location attribute has resulted in services like location-based spamming (LBS) that utilize this attribute as their coordinates for delivering services. It is important to note that while LBS has a multitude of very genuine and user-friendly applications, an offshoot of a location-based service is location-based spamming, which can cause angst both to the user and to you. (We discuss location-based spamming in more detail in the section "Infesting enterprise systems by using location-based services," earlier in this chapter.)

One of the top challenges for you is to ensure that the smartphone's geopositioning ability is protected so that the location of your employees is not inadvertently disclosed by spurious applications. You need to develop in-house tools or deploy third-party tools that guard this attribute religiously and warn the user about any potential disclosure of location to unscrupulous persons. (The most obvious culprit to use and misuse this information is a web browser that employs the Geolocation API, which is a web standard.)

Browser vendors such as Mozilla (see Figure 2-10) are giving more controls to the end user to protect this critical information, but it's unfair to burden the end user with these kinds of critical decisions. It's up to you to both educate and preferably automate these kinds of decisions on behalf of your users so their security is not compromised.

Figure 2-10:
Mozilla
location-
aware
browsing
controls.

It might sound spooky, but there is an element of your users' physical security at risk here, too. Imagine that the snooper has critical access to some enterprise assets and therefore launches a location-based theft. Some of the most popular social network applications — Foursquare, Twitter, and Facebook — have geolocation features for users to track their friends, and this information can easily be used by not-so-friendly folks for nefarious activities.

Imagine that a high-ranking individual in your finance department ends up with one of these nefarious applications on his mobile device. Based on the location information that the device advertises, the user's whereabouts can be tracked down, and the device can be pilfered to obtain access to sensitive financial information.

Mobile security 101 classes

You must take it upon yourself to educate the mobile device users who connect to your network. This training could take various forms: online webcasts, live discussions, and some real-world classes to drive home the point. It isn't easy to convince these "smart" smartphone users to take the time to understand issues that they don't even know exist, so some amount of doomsday-scenario portrayal that gets their attention is encouraged.

You might consider revoking all access to the enterprise network until the user has taken the training and passed the quiz with a respectable grade.

Because user awareness is so critical to the success of mobile device security in the enterprise, the time and effort invested in creating the training program and certifying the user community are well worth it. Trust us on this point.

We highly recommend periodic recertification of the user community because the target is always moving. The training also has to be kept up to date. Unless the solution is kept current, with periodic updates to the policy and the corresponding education of the workforce, the solution becomes impotent.

One thing you should be aware of is that the language used to educate users should be simple. Explaining the policy with anything that is overly technical, convoluted, or full of jargon will result in nonconformance or defiance. For instance, Figure 2-11 could be a poster on every users' corkboard that they can glance at every now and then to refresh their memory. The language is clear, concise, and simple.

Figure 2-11:
A sample
mobile
security
cheat sheet.

– Keep possession of your phone with you at all times
– Do not install any applications without verifying the
   credentials of the author
– Do not turn on "location tracking" by default
– Turn off any wireless interfaces you are not actively using
– Be aware of applications that store confidential data in the cloud

Turning mobile devices into allies

However daunting all this may at first seem, there is no getting around the fact that you need to have a presence on any end device connecting to your network. Watch for this catchphrase: "Connecting to the network."

Your users connect to the network with various types of devices with myriad operating systems. Preinstalling a client on all of these devices is an intractable problem, but there is a common action here that you can leverage. In essence, whenever a smartphone attaches to the network — defined here as some sort of authentication scheme followed by encryption — this mandatory action of authentication can be followed by a network-initiated device agent download, which is customized to the device attaching to the network and therefore guarantees that a client resides on every smartphone attaching to the network. Figure 2-12 illustrates how a device agent works.

Once the client has been installed, it can be viewed as a *smartphone device agent* that can then be appropriately leveraged to enforce the enterprise smartphone policy. The client provides a foothold into every smartphone attaching to the enterprise network, regardless of whether the assets are owned by the enterprise or your end user. The end user is not terribly inconvenienced because the post-authentication download is a fairly transparent process that happens automatically following successful authentication.

A device agent also allows you to actualize the impact of the constantly evolving device and application landscape and translate that into appropriate updates to this network-initiated client download so that you can stay abreast of the security hazards of this rapidly evolving market.

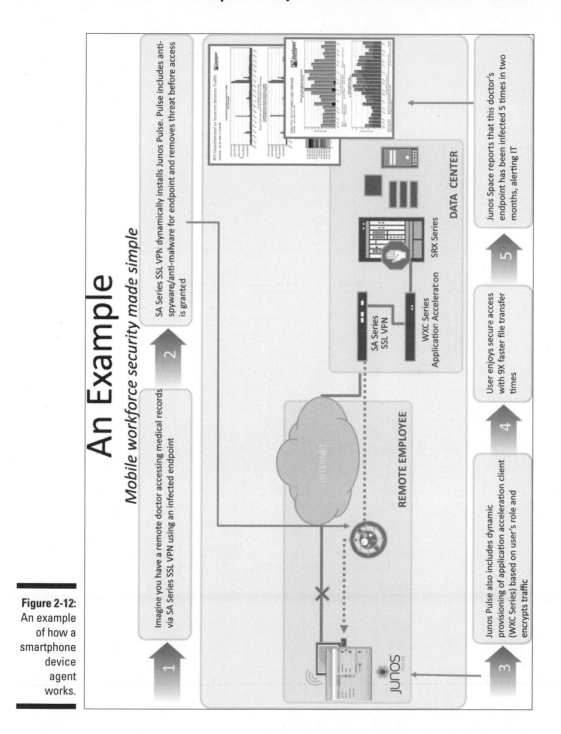

An Example

Mobile workforce security made simple

1. Imagine you have a remote doctor accessing medical records via SA Series SSL VPN using an infected endpoint

2. SA Series SSL VPN dynamically installs Junos Pulse. Pulse includes anti-spyware/anti-malware for endpoint and removes threat before access is granted

3. Junos Pulse also includes dynamic provisioning of application acceleration client (WXC Series) based on user's role and encrypts traffic

4. User enjoys secure access with 9X faster file transfer times

5. Junos Space reports that this doctor's endpoint has been infected 5 times in two months, alerting IT

REMOTE EMPLOYEE

DATA CENTER

SA Series SSL VPN

WXC Series Application Acceleration

SRX Series

Figure 2-12: An example of how a smartphone device agent works.

Chapter 3

Planning for Mobile Devices in the Enterprise

In This Chapter

▶ Protecting mobile devices from malware

▶ Remotely managing mobile device policies

▶ Enabling application access to mobile users

▶ Adapting your corporate policy for mobile devices

*M*ost corporate mobility policies allow for employee usage of one, or maybe two, approved devices for corporate use. Recent increases in the choices of mobile devices, and their increasing popularity, present challenges with today's mobility policies.

Enterprises that are used to issuing corporate-approved assets like laptop PCs are tempted to account for smartphones in the same manner, by issuing corporate-approved tablet computers and smartphones. The challenge therein is to qualify and approve these kinds of devices as quickly as they appear in the market.

Other enterprises look to migrate to a "bring-your-own-device" model, where they allow employees to bring their own devices to work, as long as the enterprise policies can be reliably deployed and enforced on these devices.

Such challenges require revisiting your existing corporate mobility policies to account for smartphones and other new mobile devices. It seems obvious that the latest generation will revamp mobility in the enterprise as we knew it. Gone are the days when employees relied solely on their corporate-standard BlackBerry device to check corporate e-mail. Today, employees have more choices for devices, many of which are not approved or, worse, evaluated by their IT departments. It therefore has become important to devise security policies for not just the corporate-approved mobile devices, but also the devices that are owned by employees and bound to be used for corporate access. These devices may include the latest gadgets available in the market.

Some devices in the market are as computationally powerful as laptop computers used to be just a few years ago. The developer platforms available for leading platforms like the iPhone and Android also facilitate a thriving ecosystem of apps, including ones that can be used for corporate access. These apps allow users to not only check e-mail but also use other applications, like client-server applications such as SAP or Oracle. This necessitates creating policies to allow users to access only those applications that they are authorized for.

The latest mobile devices are also vulnerable to viruses, malware, and other types of threats that typically are known to affect Windows PCs. This makes the security of mobile devices just as important as securing regular desktop or laptop computers. Be sure to check out Chapter 6, where we discuss the protection of mobile devices from various threats like malware, viruses and spam.

Managing the New Wave of Mobile Devices

The success of the Apple iPhone set off a trend of similar smartphones from other vendors, including Motorola, Google, LG, Samsung, Nokia, and others. Of course, the erstwhile king of the corporate phone market — the BlackBerry — still remains widely used in workplaces worldwide. Such a plethora of these phones are available in the market that competition is forcing rapid innovation from several vendors; therefore, these devices and the platforms they run on are evolving rapidly.

Just when you thought smartphones were the only hot things in the market, Apple unleashed what appears to be another game-changer: the iPad. And following suit, several other vendors either released their own tablets or announced their intentions to release tablets.

Many enterprises find that employees are abandoning their corporate-standard devices in favor of the latest and greatest new gadgets available in the market. As you can imagine, this wrecks your company's carefully laid out mobility management and security policies.

Keep in mind these two key ideas regarding the impact of today's increasing usage of smartphones:

> ✔ **Not all devices are created equal.** Smartphones are available from various device manufacturers, including Apple, Google, Nokia, Motorola, RIM, and Samsung. However, these devices are not similar in many

respects. They run different operating systems, have vastly different capabilities and features, and present their own unique challenges to being managed or secured for enterprise usage.

✔ **Smartphones are very different from regular Windows and Mac computers.** Smartphones and tablets are very different from regular computers running Windows, Mac, or Linux operating systems. Mobile operating systems like Apple iOS, Google Android, Nokia Symbian, BlackBerry and Windows Mobile are designed specifically for these smaller and more portable devices. So you'll need to give them a closer look, because your traditional enterprise policies for managing Windows and Mac systems will most likely not apply to these mobile devices.

Support the cutting-edge devices

Lots of people today are mobile enough that they don't make a distinction between work and home. Such people are technically savvy and use the newest gadgets for work as well as for personal use. These are the expert users who don't constrain themselves to using the corporate-assigned BlackBerry devices to check e-mail, but instead bring the latest available tablet or smartphone to work.

Such expert users abound in today's enterprise environment, representing the employees who buy their own mobile devices, download business apps from the market place or application store, and productively work on these devices. They like to be on the cutting edge of technology, leveraging the latest and best the industry has to offer. IT departments scramble to keep up with these users and their devices because many are simply not equipped to constantly evaluate all the latest mobile devices in the market.

Interestingly, it is difficult to ignore this kind of usage or deem the latest devices as being unsupported. Often these expert users are executives who shop for the latest gadgets and find ways to use them for corporate access. As the mobile market heats up with multiple vendors, prices become competitive, and many of these devices become affordable to mass consumers.

The bottom line is that mobile-savvy users are here to stay, and they are rapidly growing in number. Your enterprise needs to revisit your policies of handing corporate devices to employees, and analyze how you will adapt to this new trend of using personal devices for work. Here are some questions to consider:

✔ How do existing mobile security policies evolve?

Would you allow employees' personal devices into the network? How would you handle the employees who bring the latest consumer gadgets into the workplace?

✔ How do you manage these personal devices?

Would you continue assigning corporate-approved devices with custom applications and "locked-down" policies? Doing so would necessitate you to stay at the cutting edge of the smartphone market by evaluating the coolest and newest devices available in the market. This requires a lot of time and investment in both devices and personnel.

✔ How can enterprises protect themselves from losing corporate data, such as e-mail, when these devices are lost or stolen?

What type of security software would you consider deploying on these devices to protect them from viruses, malware, and other threats? There are mobile security solutions available in the market today that you need to evaluate and shortlist for deployment.

More than just e-mail

For each of the leading smartphone platforms available — such as the iOS, Android, Windows Mobile, BlackBerry, and Symbian — there are application stores supporting a variety of business apps.

Smartphone users can now easily access corporate e-mail, with sophisticated integration with Microsoft Exchange servers. They can also access web pages on the intranet using mobile browsers that support SSL encryption. Many business apps found in application stores provide functions such as Remote Desktop Protocol (RDP) or Virtual Network Computing (VNC). Several application vendors have also released client applications that enable users to access server applications in the enterprise data center.

So employees are using these devices not just to check e-mail but also to check the latest company news on the intranet, watch company videos, update blogs on the intranet, and also access server applications like SAP and Oracle. It is therefore not just corporate e-mail that ends up on modern mobile devices, but a lot more content. Mobile devices can establish *VPN (virtual private network) tunnels* (connections) to your corporate VPN gateway, thereby getting on to the network.

As devices grow more sophisticated in screen resolution and processing, this trend will only grow because application access will become ubiquitous. Although RDP on the small screen of the iPhone is cumbersome to use, it is now an order of magnitude better and more usable on the larger iPad. And if a couple of application vendors have client apps in the App Store, you can rest assured that their competitors will quickly follow them with their own apps.

Most employee mobile usage can broadly be classified into the following types of application access:

- ✔ E-mail
- ✔ Web-based applications on the mobile browser
- ✔ Full network access, including using client-server apps such as Oracle or SAP

When you think of enabling remote access for mobile devices, think of which types of applications you want to enable for access from mobile devices. In many cases, depending upon the user's *role* in the company (such as "employee," or "finance," or "IT contractor," or "executive"), a single application type or maybe two might be sufficient.

And so, it goes on. Business applications are growing rapidly in the app stores, and devices are growing more sophisticated for users to do real work on them. If everything seems to collide, take a look at Figure 3-1, which will help you visualize the challenges in the following distinct arenas:

- ✔ **Mobile device choices:** What types of devices should be allowed into the workplace and which ones should not?
- ✔ **IT enablement of new applications:** How would new applications being developed by IT be enabled access from mobile devices?
- ✔ **Mobile security:** What type of security needs to be enforced on the mobile devices, and what types of threats should they be protected from?
- ✔ **Granular access control:** What type of VPN access should be enforced on the mobile devices?

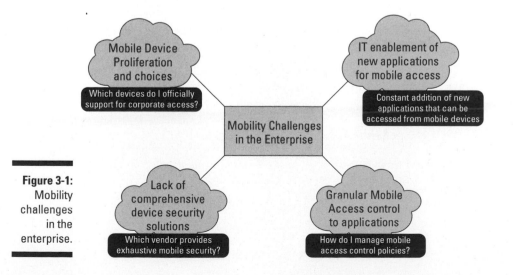

Figure 3-1: Mobility challenges in the enterprise.

The BlackBerry also supports an application store, App World, which offers a number of business apps. The BlackBerry Enterprise Server, widely deployed in enterprises, manages the deployment policies of applications on corporate BlackBerry devices. This kind of tightly controlled management model does not exist for many other popular smartphone platforms. As you begin thinking about supporting a *heterogeneous* mobile environment (an environment that contains devices from different manufacturers), you need to strategize about how you want to restrict or control applications installed on these devices.

Who moved my application?

Along with rapid mobile device innovation, there are changes happening on the application side as well. With an increasing number of applications being developed or used within the corporate workplace, the economics of cloud computing are beginning to resonate with enterprises. It has become cost-effective for many enterprises to move certain applications to the cloud, from earlier deployments on physical servers in their data centers.

It is now common to hear examples of enterprises deploying their applications in private cloud or public cloud infrastructures:

- ✔ **Private cloud:** An environment hosted within enterprise premises, but managed and operated by a different vendor, such as a service provider.

- ✔ **Public cloud:** An environment that is hosted, managed, and operated in a data center accessible to the general public. Applications such as Gmail, Google Apps, and Amazon S3 are examples of public clouds.

As applications move to the cloud, access to them is often facilitated by simple web browsers. This makes access from smartphones easier, but more challenging for the enterprise. No matter where the application is hosted, you need to secure access to it and allow access to only those users whose roles permit it. Managing access policies when you own the application and the server it runs on is relatively simple. But managing access to publicly hosted applications on employees' personal mobile devices is a different proposition.

Enforcing access control to applications has to depend upon the user's privileges and possibly change depending upon what device or location the user is connecting from. You may want to consider limiting the users' privileges to just e-mail access when they are using that latest new gadget in the market, but grant them full network access — including application access — while connecting from their corporate laptop computers.

Whenever you decide to move a certain application (such as e-mail or maybe an HR application) to the cloud, be sure to think about how this will affect access from mobile devices. For example, consider how mobile users will access the application from their smartphones or tablet devices. And think about whether you will assign different access permissions to the user, depending upon whether they are using their Windows PC to access to the application, versus their shiny new Android tablet.

Updating your mobility policies

To keep up with today's environment, you need to update your mobility policies as well as think about supporting more than one device platform. Here are some of the areas in which to consider modifying your current mobility policy and ways that you can do it:

- **Current policy:** You have a single-device policy; only one device is corporate-approved.

  Solution: Evolve to a multiplatform and multivendor policy, allowing devices of various platforms to access your corporate data. Allow more employee choice, while at the same time protecting your network. Explore mobile device management (MDM) solutions that support multiple mobile platforms, including Apple iOS, Google Android, Windows Mobile, Nokia Symbian, and BlackBerry.

- **Current policy:** Your IT department manually downloads software to each mobile device, thereby increasing the deployment costs of pushing software to mobile devices.

  Solution: Explore solutions that allow the user to download software, such as the VPN application, from an app store without needing IT to intervene.

- **Current policy:** You deploy endpoint security only for Windows.

  Solution: This is no longer a sufficient security solution. Look for solutions that can protect mobile devices from malware, viruses, and other threats. For more information, be sure to read Chapter 6, which describes the types of threats from which you should protect these devices, and the solutions that offer the appropriate form of protection.

- **Current policy:** You have application policies in place only for Windows.

  Solution: Most enterprises have systems in place that can deploy applications to Windows PCs. You need to *scale* (adapt) this ability to mobile devices as well, to manage and control the apps that are installed on them. Explore solutions that give the ability to restrict apps on mobile devices to a list that you can manage.

✔ **Current policy:** You have no loss and theft-prevention policy for mobile devices.

Solution: Mobile users are vulnerable to losing valuable data on mobile devices when those devices are lost or stolen. Many enterprises lack the policies to mitigate the risk of losing sensitive data on such devices. Look for solutions that allow you to take immediate preventive action on lost or stolen devices. Actions include remotely locating a device via GPS, remotely setting off an alarm, or remotely wiping selective device contents, such as personal data or corporate data, or both.

Adapting to the New Challenges of Mobile Devices

Most enterprises today have designed mobility policies centered around the usage of just one type of corporate device, which in many cases happens to be the BlackBerry. Some have recently adapted to include popular Apple devices in their corporate policy as well. As you ponder migrating from this model to a more flexible one, supporting many of the latest devices, here are the aspects of mobility policies you should revisit:

✔ **Protecting mobile devices from malware and viruses:** This should be the most fundamental requirement for allowing any device to access e-mails, applications, and data on your corporate network. No device running any kind of malware should be allowed onto your corporate network. Protecting the devices is paramount to protecting your own network from attacks originating from such devices.

Part of preparing your network for mobile devices is having the means to protect *all* the mobile devices on your network from malware, viruses, and other threats. Key items here include installing *endpoint security software* (software to protect the network when accessed remotely), ensuring that the software remains updated with the latest virus signatures, and ensuring that you can deploy this on the latest devices hitting the market.

✔ **Remotely controlling device security policies:** The ability to remotely set device policies (including password policies, inactivity timers, application policies, and so on) enables you to control and change device policies from a central application on your network.

Imagine you log on to a management console within your network and set a new password policy, requiring users to have a certain password strength. This policy should then propagate to all the connected mobile devices in your network, requiring all users to conform to your new password policy.

Other items in this area include planning for backing up and restoring corporate data on the device, encrypting sensitive content, controlling the types of applications installed on the device, and taking action when the device is lost or stolen.

✔ **Enforcing granular access control for users connecting from mobile devices:** After you've put a plan in place to secure mobile devices, think about what types of applications you want users to be able to access from their mobile devices, and for what groups or users. For example, you should grant application access only to those users who absolutely need access to those applications. _Granular access control_ refers to policies that control access to finance-related applications only to your corporate finance group, and HR applications only to the HR group, and so on. Managing and enforcing granular access to applications and data are critical for a successful mobile policy implementation.

Sophisticated smartphones can be used to access corporate e-mail, web-based applications on browsers such as the intranet, or even client-server applications such as Oracle or SAP. You might want to consider which of these applications you are willing to enable on mobile devices, and for which users.

This also includes enforcing strong authentication and authorization policies for users when they log in from mobile devices. Ideally, you want to enforce the same authentication methods from mobile devices, including _multifactor authentication_ (authentication based on two or more factors), as you do on regular laptop or desktop computers.

We discuss these areas of your mobility policy in more detail in the sections that follow.

Protecting mobile devices from malware

Applications are deployed to mobile devices such as tablets and smartphones from respective app stores or markets. The number of app downloads from such markets now ranges in the billions on smartphones worldwide.

When your company's employees use third-party apps for applications like online banking, checking corporate e-mail, and playing interactive games with their friends, these devices become appealing targets for hackers. Because many of these devices run relatively new operating systems like Android, Symbian, and iOS, hackers fancy their chances of exploiting platform vulnerabilities to steal information from these devices.

Market research shows a rapidly increasing occurrence of mobile malware from 2008 to 2010. This increase is in line with the corresponding rapid increase in device options available in the market. So, if you are going to allow smartphones to access your critical corporate data, it would be prudent to plan for a scalable (adaptable) and reliable way of protecting them and your network from malware.

There are two broad options available for providing threat protection on smartphones and other mobile devices:

✔ Client-based mobile endpoint security software

✔ Cloud-based software

We discuss these options in more detail in the following sections.

Client-based mobile endpoint security software

In this type of deployment, an actual client software app protects the device from viruses, malware, spam, and other threats. This is similar to how client endpoint software is deployed on regular Windows computers. Software available for mobile devices is usually designed to run in the background, scan the device periodically for threats, and *introspect* (analyze) data received on the device for viruses and malware. Such software typically alerts the user when a threat is detected, and automatically quarantines or deletes the source of the threat as well. Symantec, Trend Micro, F-Secure, McAfee, and Juniper offer client software–based mobile security solutions.

Software applications are typically deployed to mobile devices via the following two ways:

✔ Downloaded via the app store by the users themselves or deployed via a mobile device management system by the IT department.

✔ Deployed automatically over the air (OTA) from a server that the device connects to. This approach typically happens with no user intervention.

Virus signatures are typically updated in a central system periodically. Then devices either *download* the signatures at regular intervals or they're *pushed* out to devices periodically.

With client-based software, there are some basic things to watch out for while shopping for a suitable mobile endpoint security solution:

✔ **Determine what device resources are used by the software**. You certainly do not want to deploy client software that drags down the performance of the device. So look for the following attributes while narrowing your options:

• *Size of the client software:* Needless to say, the smaller the client, the better.

- *CPU utilization:* The software should run as unobtrusively as possible, reducing any impact on the user's activity on the device. If running the application slows down the entire device, then it is apparent that the application is taking up a lot of system resources to function.

- *Memory utilization:* The software should consume as little memory as possible. Again, like the impact on CPU utilization, when an application consumes too much memory, it drags down the performance of the device in general.

✔ **Avoid software that is ported to a mobile platform from Windows**. Beware of software options that are essentially desktop endpoint software ported for the mobile platforms. *Porting* in the software development world refers to the process of customizing software for a different platform than what the software was initially designed for.

Several vendors offer endpoint security software for Windows platforms. When you shop for mobile endpoint security software, make sure that the mobile endpoint software was designed from the ground up for each specific mobile platform.

Investigate options that allow for simple deployment of the software to mobile devices. You don't want the IT department to have to deploy the software manually to every mobile device used by employees. A simple deployment mechanism like OTA or availability in the app store is probably most desirable.

Cloud-based security

In this type of deployment, the actual threat protection happens in the cloud or centralized data center of the endpoint software vendor. Traffic to and from the mobile devices is redirected on the device to the cloud for malware detection.

Typically, this option includes no client-side software and relies instead on each application to take appropriate action when a threat is detected. For example, content downloaded from websites is inspected in the cloud before it's delivered to the mobile device browser. If a threat is detected in the web content, the cloud service indicates so, and the browser displays an appropriate message to the user. Zscaler, ScanSafe (now owned by Cisco), Symantec, and McAfee offer cloud-based mobile security solutions.

If this cloud model of endpoint security is what you need, make sure you analyze the following aspects of the solution:

✔ **Security between the mobile device and the cloud service:** If both Internet traffic and corporate traffic (such as e-mail and intranet browsing) are sent to the cloud, you should make sure that the traffic is flowing over a secure tunnel. You don't want anyone sniffing on the traffic that may carry sensitive data. Be sure to check with the cloud service vendor regarding the security between the mobile device and the cloud service.

> ✔ **Latency introduced by the cloud service:** If data sent and received by the device hits the cloud service before heading to its destination, make sure that the cloud service is rapid in its response. Otherwise, the user experience on the mobile device will be adversely affected. The latency is apparent from the user experience when the cloud service is enabled, compared to the situation when it is not enabled. If the cloud service adds a lot of latency, then the user's browsing and other application access are slower.

Many cloud-based solutions offer protection against web-based threats for information accessed via web browsers. Mobile devices, however, are not only vulnerable to threats via web browsers but are also susceptible to receiving malicious content via MMS, SMS, or e-mail. Be sure to investigate options that provide holistic device protection for your employees' mobile devices.

Managing device policies remotely

Now that you've thought about securing the mobile devices on your network from threats like viruses and malware, it's time to plan for remotely enforcing policies for device management or security.

No matter how powerful the endpoint security software is on the device, the following types of user behavior pose direct risks of losing valuable data on the device:

> ✔ Not locking the device
>
> ✔ Not setting a secure password (for example, having "1234" or "abc123" for a password!)
>
> ✔ Storing passwords in third-party apps, such as online banking apps or an Oracle app that can directly access the latest sales pipeline

You get the idea. We're talking about device security etiquette, about taking the simple yet often-ignored steps for protecting vital data stored on the device.

There are two broad categories of actions you will need to take on mobile devices in your network:

> ✔ **Mobile device management:** Remotely managing the devices, including enforcing the need for a passcode or deploying a set of corporate-approved mobile apps to them.
>
> ✔ **Remote device security:** Remotely securing the devices, including taking preventive action when the devices are lost or stolen.

Configuration and application management

Similar to how you deploy software systems to manage desktop and laptop computers, you need to think of software that can manage the diversity of mobile devices available in the market.

For example, the BlackBerry Enterprise Server is an excellent candidate, but it falls short on one major area: It is a solution only for the BlackBerry. It doesn't help manage other types of devices, such as those running iOS, Android, or Windows Mobile.

Remote device management policies typically include configuration management and application management as follows:

- **Configuration management:** Involves deploying IT-approved software versions of supported mobile platforms. It is ideal to find a single solution that can manage the configuration for a heterogeneous mix of mobile devices. If you cannot find a single solution that can do so, try to minimize the number of systems you would need to deploy. Configuration management includes things like managing the OS version of mobile devices and application and security patches, or supporting any other desired corporate policy.

- **Application management:** Involves controlling the apps deployed on mobile devices. If you're worried about mobile devices on your network running apps that you've never heard of or apps that are known to be insecure, plan for deploying application control policies to those devices. Such policies include viewing an inventory of all applications installed on devices in your network and being able to view the details of each application and the devices running it. You should also be able to select a particular application and either uninstall it from users' devices or send messages to users that those applications are not corporate-approved and must be uninstalled.

  This is similar to certifying certain applications as safe applications, depending upon the criteria of your choice. This could enable you, for example, to deem certain apps forbidden within your network, or restrict all mobile apps to a predefined list you come up with. If you desire some level of application enforcement and control, be sure to evaluate vendors that can restrict applications installed on mobile devices to a predefined set.

- **Backup and restore:** Make sure you think of a way that you can back up contents of mobile devices running in your network. This is as important as backing up contents of desktop and laptop computers. Having a sound system in place for this critical function could make a great difference in improving the productivity of mobile device users, who should be able to replace devices easily if you back up their data.

This function enables backing up data from employees' mobile devices and allows seamless restoration of data, potentially to a replacement device running a different mobile platform. If your enterprise IT can do so, it's a valuable service for employees, as well as an assurance to you that users will be productive immediately after moving from one device to another. This is like replacing laptop computers for users, with their data restored immediately to the new laptop.

Chapter 12 describes the backup and restore policies in more detail, including the selective backup of certain content on corporate-issued versus employee-owned devices.

Security of lost and stolen devices

There are various actions you need to take when an employee reports a lost or stolen device. You should be able to do the following:

- ✓ **Remote lock:** Remotely lock the device so nobody can log in to it.
- ✓ **Remote alarm:** Remotely set off an alarm so that the device makes itself heard!
- ✓ **Remote location:** Remotely find the device using its GPS capabilities.
- ✓ **Remote wipe:** If all else fails, and if you are sure the device is lost, you should be able to wipe the device clean of all or selective data.

So protection against loss and theft is an example of securing devices remotely when corporate data is at risk of being lost on them. Other types of security policies include setting password policies, such as the required strength of the password, or setting an inactivity timer to automatically lock the device.

Even after deploying a best-of-breed security solution for mobile devices, make sure that employee carelessness does not become the weakest link in your security implementation. Be sure to set password policies requiring a password on every mobile device, and impose an inactivity timer on every mobile device. Doing so prevents the leaking of corporate data via eavesdropping or other means when mobile devices are not sufficiently secured.

Enforcing granular access control

If you've taken the advice we give earlier in the chapter, you've begun to devise a plan to secure mobile devices from malware and viruses, and you've also planned on managing these devices remotely, including being able to remotely wipe or lock them. The third step of ensuring that these devices are corporate-ready is to enforce granular access control on users connecting from these devices.

You may not want to enable all mobile device users to have the same access privileges as they do on their regular Windows or Mac computers. For example, you may not want all mobile users to have full network access, including access to your corporate customer relationship management application that tracks the latest sales deals. You may want to enable only access to e-mails or certain web-based applications to some groups of users.

Another key item of access control is the authentication policy itself. To allow users to access corporate stuff from mobile devices, you should not relax any security policies of enforcing strong authentication on these users. You should think of the authentication methods to enforce on mobile device users and the backend systems you will need in place accordingly.

Finally, the key to a scalable mobile infrastructure is to have a single place of managing all your policies of access control, authentication, and policy enforcement. If you have a VPN solution in place, you probably already have policies in place that control the access to applications to specific groups of users. This is the key piece that binds it all together: How should you leverage the policies on such a centralized VPN system to mobile security?

The following sections discuss the key elements of implementing a flexible access control solution for mobile devices.

Authenticating users

The most fundamental requirement to allowing mobile devices within the enterprise is to have a solution in place to authenticate the users of those devices. It is common to use the following methods to authenticate mobile device users:

- ✔ Authenticate using username and password.
- ✔ Authenticate using a certificate deployed to the mobile device.
- ✔ Authenticate using one-time passwords or security tokens. One-time passwords expire after a single usage, thereby preventing hackers from attempting to use a password after it has already been used once. Such passwords are usually deployed using *tokens,* either hardware dongles from vendors like RSA or software applications that issue a unique password every time.
- ✔ Authenticate using smart cards.

Many enterprises implement *dual-factor* or *multifactor authentication systems,* which means that multiple authentication methods are cascaded one after the other, to enforce strong authentication. For example, a user may be prompted to authenticate using her username and password, and then prompted again to authenticate using her one-time password and PIN.

Ideally, you want to leverage the same authentication infrastructure to authenticate mobile devices as for regular Windows, Mac, or Linux systems. For example, if you've already deployed RSA SecureID two-factor authentication for regular desktop and laptop systems, enforce the same level of security on mobile devices as well. This will save you time, money, and hassles.

If you need to enforce certificate authentication on mobile devices, you need to look for management solutions that can deploy certificates to devices at scale. Look for such capabilities in the management systems you already have in place for deploying certificates to Windows PCs, for example. Several existing management solutions have recently added mobile features to manage certificate deployments on all types of devices.

Authorizing users to see only the data they are allowed to see

Once users authenticate successfully from mobile devices, allow them to access only the data or applications that you want them to. You may not want all users to be able to access any or all types of applications by default. Many mobile device users want access to only corporate e-mail, whereas others use these devices to check the intranet web pages. Yet another type of users, power users, want to log in to their remote desktops and remotely operate their desktop applications from their mobile devices.

Here is a broad categorization of application types that you may want to restrict access from or allow access to, depending on the group that a user belongs to:

- **Web-based applications:** Users can access intranet pages from mobile device browsers.

- **E-mail:** Users can send and receive e-mail and schedule meetings on the calendar.

- **Full network access:** Users can access not only web-based apps and e-mail but also any other corporate client apps on the mobile device downloaded from an app store.

You can allow mobile users to access web-based applications and e-mail without letting those devices into the corporate network, such as by assigning them an IP address within the network. Web-based applications can be accessed by most sophisticated mobile browsers supporting SSL encryption. E-mail access can be enabled via Microsoft Exchange or ActiveSync, which also does not need the mobile device to have an IP address within the network. Full network access, on the other hand, needs the device to be within the corporate network. This type of access allows the user to access pretty much any application within the network, just as if they were in the office. Accordingly, your security policies need to be at their strictest for granting full network access.

Integrating with existing VPN policy infrastructure

If you allow your users VPN access to the corporate network, you likely already have a policy in place that describes what types of users are allowed access, including the applications that are allowed to be accessed remotely. VPN policies are typically enforced on a VPN gateway device at the perimeter of the network, with access for external users.

While shopping for VPN solutions for mobile devices, look for the following:

✔ Wide range of supported mobile platforms for corporate access, such as these:

- *Apple iOS*
- *Google Android*
- *Windows Mobile and Windows Phone 7*
- *Nokia Symbian*
- *BlackBerry OS*
- *Others such as HP Web OS*

✔ Wide range of supported authentication methods:

- *Username and password–based*
- *Certificate-based*
- *Multifactor authentication (for example, cascading username and password-based authentication followed by certificate-based authentication, or vice versa)*
- *VPN on demand (setting up a VPN tunnel automatically when the user attempts to access a corporate resource)*

✔ Ability to assign role-based access to users, depending on their role within the enterprise

✔ Ability to assign granular access to any or all of the following types of applications:

- *Web-based intranet content*
- *E-mail*

✔ Full network access

VPN gateways are typically either dedicated VPN appliances that enable IPsec or SSL VPN access, or firewall devices that include VPN functionality in addition to a host of other security features. In either case, most VPN solutions should have a well-defined policy infrastructure to define role-based access to corporate data and applications.

Depending upon your corporate policy and need for application control, you should choose between an IPsec VPN solution or an SSL VPN solution. Here is some information that can help you choose between the two:

- **IPsec VPN solutions:** Enable full network access to remote users. That means users who connect over traditional IPsec VPN tunnels are granted full network access to the corporate network, including getting an IP address within the network.

- **SSL VPN solutions:** Usually allow more granular access control, enabling you to control application access to any of all of the various application types: web-based, e-mail, or full network access.

Choose a solution that allows you to manage mobile access control policies on this kind of a centralized VPN system that already manages remote access policies. It would be counterproductive (and very costly!) to manage duplicate or redundant policy systems, one for traditional remote access from home PCs and another for mobile devices.

To integrate your existing VPN policies with mobile access control, here are the key decision areas you need to consider:

- **Your mobile security solution:** Depending upon what security features you need on your users' mobile devices, choose a solution that spans a broad range of mobile platforms. As discussed earlier, you may choose any or all of the security features to enforce on mobile devices, including protection against viruses, malware, Trojans, and spam.

- **Your endpoint security *posture* (level of risk):** You may already have an endpoint security solution on your VPN gateway allowing network access only to devices that have a sufficient security posture. This policy may include checking for installed antivirus or antimalware software, or verifying that the device is a corporate-assigned computer before granting VPN access. You may want to extend this policy to mobile devices, allowing VPN access only to those mobile devices that are secured by the security software of your choice.

- **Your access control policies:** The access control policies enforced on the VPN appliance should *follow the user,* meaning that no matter where the user logs in, the policies applicable to that user must be enforced. Choose a VPN solution that can enforce a single set of access control policies, irrespective of where users connect from, or what devices they use to connect. Having a single set of policies that span across device and application types will make your life simpler.

Integration of mobile security functionality with your existing VPN solution has several advantages, such as the following:

✔ Easy enforcement of mobile device security as an endpoint posture assessment check, prior to granting VPN access to users

✔ Easy enforcement of access control policies that are already defined on the VPN gateway

✔ Easy integration into the management capabilities of the VPN solution, thereby offering insights into the mobile device inventory and assets within the enterprise

Part II

Implementing Enterprise Mobile Security

The 5th Wave   By Rich Tennant

"We take network security here very seriously."

In this part . . .

You've just finished taking down all those No Mobiles Permitted on Site signs, not because you secretly don't agree but because no one was paying attention to them.

But luckily, Part II helps you put together a plan for mobile device security. Chapter 4 gets you on the road to recovery by helping you create policies, those glowing rock star plans that architect your security structure. Chapter 5 continues the euphoria by outlining how to manage and monitor the policies you implement. It's not rocket science, and chances are you've already implemented many of them — but who knows. Maybe we are actually rocket scientists, minus the rocket scientist paycheck.

Chapter 6 is the real trick: making sure your plans conform to existing corporate compliance policies so there's a united voice to the user about future compliance.

And as a mobile security bonus, each chapter in this part has a case study at the end, explaining how our book's model company, AcmeGizmo, implements the concepts discussed in that chapter.

Chapter 4

Creating Mobile Device Security Policies

*T*his chapter delves into the universe of security policies for mobile devices — the why, how, what, and where. Clearly the danger wrought by these new age devices should be apparent by now. In this chapter, we shift the focus to how you can combat this danger with consistent, transparent, and comprehensive policies that both protect the enterprise and educate your users.

Recognizing the Importance of Enforceable Security Policies

Before we get into the nitty-gritty of the various components of security policies, it is important to understand the need for them. If every one of your users were an intelligent, security-savvy, self-regulated, and enterprise law-abiding citizen, you could do away with enforcing the policies altogether. The only aspect of the policies that you would need to worry about would be the creation and education pieces. However as you know, life is not cut-and-dry, and your users are typically very innovative when it comes to skirting the rules, not prone to reading policy documents or understanding the impact of noncompliance, frequently try to circumvent the policies that exist, and constantly excel in their ability to figure out loopholes.

Therefore, your security policies need to include the following:

✔ Unambiguous terms and definitions that are universally understood

✔ Language that enables enterprise IT — you — to codify the rules of engagement so that both you and your users can adhere to an unambiguous set of documents

✔ In the event of a breach of policy, the ability to take remedial action with the primary aim of protecting the enterprise and, in the event of violation due to nefarious intent, to follow prescribed guidelines against the errant individual

✔ The ability to adjust the policies based on user feedback and deployment-related learning

Figure 4-1 depicts the five stages that an IT policy lifecycle passes through, and this is applicable to a mobile device security policy as well. Here is a brief description of the five phases shown in the figure:

✔ **Define the policy.** This stage stipulates the policy in clear and concise terms.

✔ **Educate the users.** In this stage, it's critical that you clearly communicate the policy to the users. Make sure you get your message across.

✔ **Implement the policy.** This stage sets into motion the actual policy itself.

✔ **Audit the policy.** This is the data collection and feedback stage to assess how the policy is performing versus its stated objectives.

✔ **Modify the policy.** This is a crucial but often overlooked step: to be able to adjust the policy based on the results of the audit and the feedback gathered.

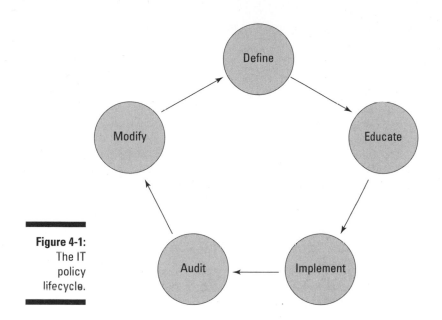

Figure 4-1:
The IT
policy
lifecycle.

Understanding Device Policies

Device policies can be split into two categories: policies for approved devices and policies for other devices, as shown in Figure 4-2.

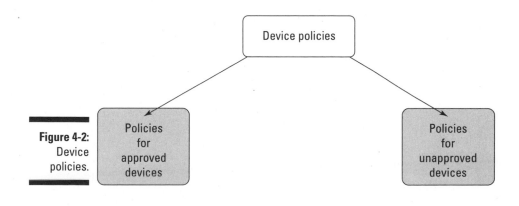

Figure 4-2:
Device
policies.

Here is a rundown of the two categories of device policies that you need to communicate to users:

- **Policy for approved devices:** This policy applies to all enterprise-issued mobile devices. Because these are enterprise assets, you are at liberty to set a strict usage policy as well as establish stringent penalties for misuse.

- **Policy for unapproved devices:** An *unapproved* device in this case is a device that the enterprise neither endorses nor supports. This does not mean that you can summarily deny all connectivity to enterprise assets, but you can impose restrictions on what, how, and when these devices connect to enterprise resources.

Obviously, your policies will be largely applicable to the approved-devices list, because this is what will typically be the exposure that your employees are subject to.

There is going to be a rapid transition of devices from the unapproved list to the approved list based on user adoption of evolving mobile devices, so expect the list of approved devices to continue to grow. For instance, when the first iPhone was introduced in 2008, there was very little enterprise IT support for it. Fast-forward to today, and a large number of enterprises (a number that is ever increasing) support this device.

The unapproved devices policy will simply be one of two options, as shown in Figure 4-3:

- **Access denied:** No access to the enterprise network altogether
- **Access restricted:** A highly constrained set of privileges available to the user

The following are the key elements to consider when creating policies for *approved* devices. Note that there is further categorization in the approved device category: employee owned and corporate issued, as shown in Figure 4-4. For each of the policies that follow, these will be called where appropriate:

- Policies for physical device protection
- Policies for device backup and restore
- Policies for device provisioning

We examine each of these policies in turn in the sections that follow.

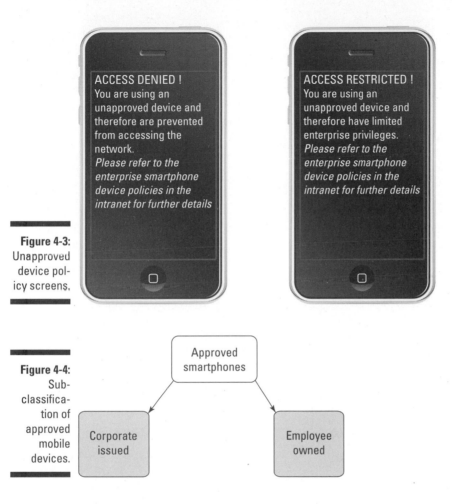

Figure 4-3: Unapproved device policy screens.

ACCESS DENIED !
You are using an unapproved device and therefore are prevented from accessing the network.
Please refer to the enterprise smartphone device policies in the intranet for further details

ACCESS RESTRICTED !
You are using an unapproved device and therefore have limited enterprise privileges.
Please refer to the enterprise smartphone device policies in the intranet for further details

Figure 4-4: Sub-classification of approved mobile devices.

Approved smartphones

Corporate issued

Employee owned

Policies for physical device protection

The policies for physical device protection are mostly common sense — and yes, how uncommon is that? Yet these concepts bear repeating because your users take a lot of this for granted, and laying out the do's and don'ts drives home the point.

Here the key tenets of physical device security (outlined in Figure 4-5) that you would convey to mobile device users at your company:

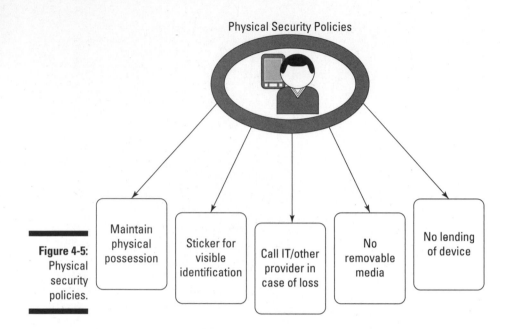

Physical Security Policies

Maintain physical possession

Sticker for visible identification

Call IT/other provider in case of loss

No removable media

No lending of device

Figure 4-5: Physical security policies.

✔ Ensure that your device is within your control at all times.

✔ Ensure that removable media usage is avoided altogether and, if that isn't possible, ensure that the data on the media is encrypted.

✔ Refrain from lending your device to third parties.

✔ Use a sticker (it's low tech, but it works) that contains your name and contact information and stick it on your device so that in the event the device is lost, there is an opportunity for a Good Samaritan to contact you.

[Create these stickers beforehand and hand them out to your users during the training process.]

✔ In the event of theft of your device, immediately contact the appropriate party. If it is a corporate-issued device, IT can initiate remote recovery and remedial operations.

✔ If it is your personal device and you have remote recovery services from your provider or device manufacturer, follow that procedure right away. In the event that you don't have any such recovery mechanisms, contact your service provider so that at the very least they can immobilize use of the device itself.

Remote recover and remedial operations

Remote recovery and remedial operations are essential functions provided by most device manufacturers and mobile operating systems vendors as well as third parties. Under the category of mobile device management, *remote recovery* entails locating the device, initiating remote wipe operations, and locking down the device to immobilize it to prevent unauthorized usage. *Remedial operations* entail locating a substitute device, restoring the state of the original device onto the replacement, and issuing the replacement to the user.

Policies for device backup and restore

As mobile devices become an integral part of the digital communication toolbox for accessing enterprise assets, and with the growing storage capacities on these devices, there is going to be an ever-increasing propensity for enterprise data or intellectual property to reside on these devices. Couple this trend with the storage of business contact information and images and videos that are captured by the mobile devices, and the need to be able to back up these devices becomes paramount. Needless to say (but we will, anyway), the other side of the equation is to be able to quickly restore these devices to an operational state using previous backups. Both these critical tasks of backup and restore are your responsibility. To make this process as painless and automatic as possible for both you and your users, you need to establish a set of policies that should be adhered to religiously.

You should look at backup and restore policies from these two viewpoints, as shown in Figure 4-6:

✔ Recommended policies for user-owned devices for backup and restore

✔ Mandated policies for enterprise-issued devices for backup and restore

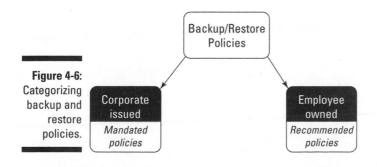

Figure 4-6:
Categorizing backup and restore policies.

This distinction is critical to the success and scalability of the backup and restore plan you put in place. You are obviously keenly aware of the proliferation of the variety of mobile devices in the enterprise and the need for you to support all of them, and therefore, the need for recommended backup and restore policies for your employees when they bring their own devices into the enterprise.

Here are the key tenets to pass on to mobile device users in your enterprise for employee-owned device backup and restore:

- ✔ Schedule periodic backups of your mobile device with your desktop, laptop, and/or remote servers.

- ✔ For extended removable media, such as SD cards, ensure that these are backed up separately (in case your device backup software doesn't do this automatically).

- ✔ When the device needs to be restored to a previous known configuration, identify a known previous backup and initiate the restore procedure. This backup might be locally stored in your data center or with an outsourced service such as Carbonite.

As is evident, these are very generic policies, and the aim for you is to provide them as a rule of thumb. Your guidance can prompt your employees to capitalize on any additional bells and whistles that a particular device or OS vendor may provide.

For instance, Figure 4-7 shows a screenshot of the BlackBerry Protect application, which allows BlackBerry users to set backup options, including using the network interface they are on to decide whether to back up. Some users may be sensitive to the amount of data traversing the network during a backup operation, especially if they're on a capped data-usage plan; therefore, the choice of backing up when the mobile device is on a wireless LAN and not backing up when the mobile device is connected to a 3G network, for instance, is very useful.

For the iPhone, the backup and restore application is built into iTunes. The user has little to do besides plug the iPhone into the computer regularly, and the backup happens automatically. Restore is also a very straightforward operation using iTunes. The user can open iTunes and select the Restore from Backup option shown in Figure 4-8.

For the Android, users have several options, including apps, service provider services, and connecting to a computer and working with folders in Windows Explorer. Also, some of the data (Gmail and other Google stuff) is automatically backed up in the cloud.

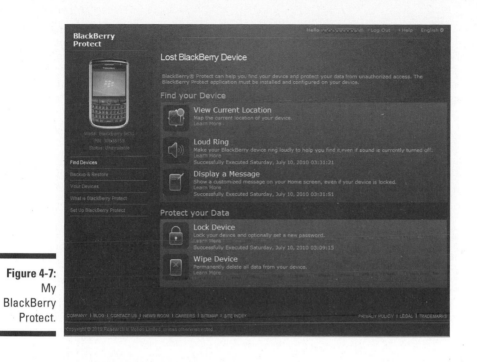

Figure 4-7:
My
BlackBerry
Protect.

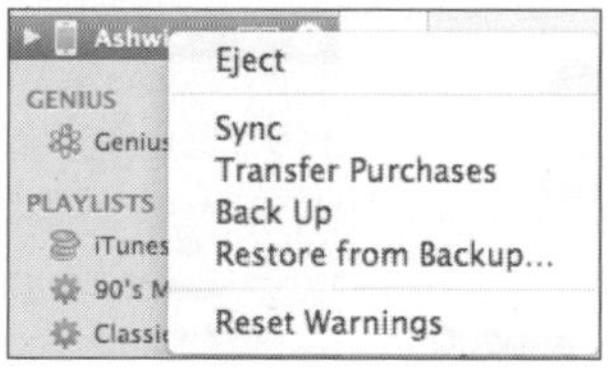

Figure 4-8:
iPhone
backup
and restore
capabilities.

Here are the key tenets for corporate-issued mobile device backup and restore policies that you would convey to users at your company:

- ✔ The data on your mobile device is automatically backed up when you connect to the network. This data includes any personal information you may store on the device.

- ✔ Tampering or interfering with this backup may result in confiscation of the mobile device.

- ✔ Notify the IT department right away if your device is lost or stolen.

> ✔ The use of removable media is highly discouraged. As the name suggests, the media can be removed and, therefore, cannot be guaranteed to be backed up.
>
> ✔ Disabling or crippling the mobile device backup agent running on the device is prohibited and may result in confiscation of the mobile device.

For disaster recovery and multilevel backup of the storage servers that are used for mobile device data backup, the mobile device backup archives should be treated equivalently to the existing laptop and desktop backup server policies you have in place. You should not have to do anything significantly different as it relates to accommodating mobile device backups. In fact, the IT analysts at Enterprise Strategy Group conducted a survey in 2010 (shown in Figure 4-9), and found that the ability to integrate with existing backup systems for data protection ranked as the 10th most important consideration among IT executives when it came to evaluating, selecting, and implementing mobile security and management solutions.

How would you rate the importance of the following features and/or capabilities when it comes to evaluating, selecting, and implementing mobile security and management technology solutions? (Percent of respondents, N=174)

Figure 4-9:
Mobile device security priorities.

Source: Enterprise Strategy Group, 2010.

Using Provisioning Policies to Manage Devices

Provisioning policies are policies that define the lifecycle of mobile device management, including setting up the e-mail configuration, configuring backup and restore, activating the device with the service provider, maintaining compliance with enterprise guidelines, tracking inventory in real time, and decommissioning the device. Keep in mind that provisioning policies are predominantly applicable only to enterprise-issued mobile devices because these policies are about giving you control of the mobile device to do the management that you need to do.

For employee-owned mobile devices, there is no explicit provisioning that you need to be concerned about. Because the employee owns the device, the onus of activating, backing up and restoring, tracking a lost device, and so on rests with the user. However, you will need to provide a well-defined set of profile settings (network, e-mail, web, and security) that users can then use to self-provision their devices for connecting to the enterprise network.

Provisioning a mobile device typically involves the following three steps, as outlined in Figure 4-10 and discussed in more detail in the sections that follow:

1. Upgrade, downgrade, and install software.

2. Upgrade profile settings.

3. Decommission the mobile device.

Upgrade, downgrade, and software installation policies

Essentially, you use provisioning policies to change the software on mobile devices. And by the disruptive nature of these changes, you (luckily) do not have to deal with this on a frequent basis. When communicating with your users, you need to note that the software on their phones is subject to periodic upgrade and downgrade, and they need to know the following basic tenets of the policies:

✔ As with any software update, there is a possibility that this may affect previous settings or cause stability issues; therefore, we highly recommend that you back up your phone prior to the upgrade or downgrade.

[Note that you are responsible for ensuring periodic backups of users' mobile devices; however, it behooves your users to assume some responsibility toward this critical task, so we highly recommend reminding users to do this before a software upgrade.]

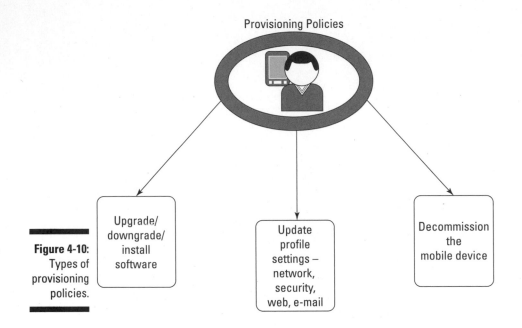

Provisioning Policies

Upgrade/
downgrade/
install
software

Update
profile
settings —
network,
security,
web, e-mail

Decommission
the
mobile device

Figure 4-10:
Types of
provisioning
policies.

✔ In the case of manual upgrade or downgrade, we will notify you through e-mail, SMS (text message), or other means that an upgrade or downgrade is available and recommended. Please follow the steps outlined in that message to comply with the software change.

If we find that you haven't performed the manual upgrade or downgrade after a period of # days *[plug in your favorite tolerance window for your users in place of #]* from when the notice was issued, your right to connect to the enterprise network may be curtailed.

✔ In the case of automatic upgrade or downgrade, you will see a pop-up message when you connect to the network, and we recommend that you back up your device before proceeding with the change.

As in the case of manual upgrade or downgrade, if we find that you haven't allowed the upgrade or downgrade after a period of # days *[plug in your preferred tolerance window for your users in place of #]* from when the pop-up message was posed, your right to connect to the enterprise network may be curtailed.

✔ The IT department reserves the right to install new software at any time to your mobile device.

Similar to software upgrades or downgrades, additional software installation could be *manual* (you need to actively seek this software) or *automatic* (this will be pushed to you).

✔ If we find that you haven't performed the software installation after a period of # days *[plug in your preferred tolerance window for your users in place of #]* from when the new software notice was issued, your right to connect to the enterprise network may be curtailed.

✔ Installation of unapproved software on the mobile device is not permitted. If such unapproved software is discovered, you will see a warning; after a period of # days *[plug in your preferred tolerance window for your users instead of #]* from when the warning was issued, if you have not removed the unapproved software, your right to connect to the enterprise network may be curtailed.

Profile settings policies

The profile settings policies that we discuss here are fundamental configuration settings that need to be provisioned on the mobile devices in order to get them to function per enterprise guidelines. Typically, these refer to web, e-mail, network, and generic security settings on the mobile devices.

Akin to the backup and restore policies that we discuss earlier in the chapter, profile settings policies are broadly classified into employee-owned and corporate-issued profile policies. The fundamental difference between the two — stating the obvious here — is that the former is the onus of the employee to configure based on the settings you provide, while the latter is provisioned by you before the device is handed to the employee.

The employee configuration of profiles isn't as onerous as it sounds. In fact, the leading mobile device vendors have made this process very intuitive, and as long as you provide employees with the correct parameters, the configuration is a very straightforward task.

The following list describes some of the configuration profiles that you can create with the iPhone:

✔ The iPhone e-mail configuration profile, shown in Figure 4-11, is fairly straightforward, and an average user would be able to configure it with ease using the relevant pieces of information: mail server address, port, username, and password.

✔ The iPhone passcode profile, shown in Figure 4-12, is a tool that you can use effectively to mandate strict passcode parameters, using alphanumeric values as opposed to numeric only, longer passcode lengths, quicker passcode ageing, auto-lock with inactivity detect, and so on.

✔ The iPhone VPN profile, shown in Figure 4-13, again follows typical industry terminology, and any users who have configured VPN on their laptops for home use should be able to follow the same logic easily on their iPhones.

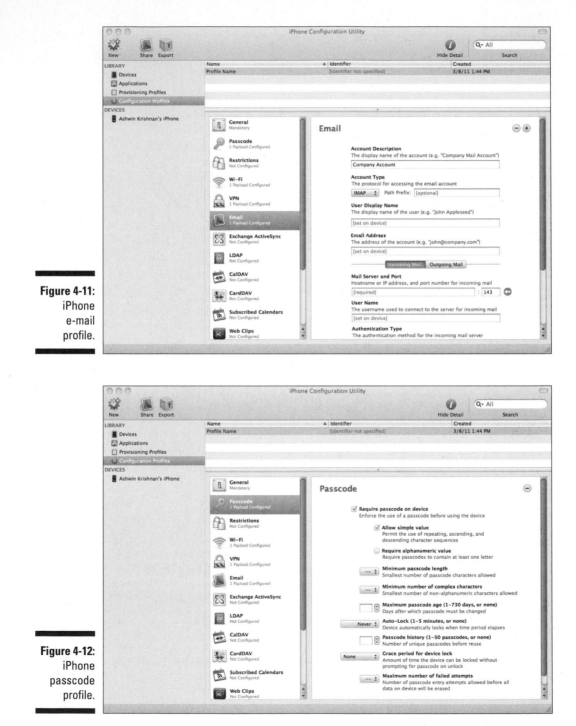

Figure 4-11:
iPhone
e-mail
profile.

Figure 4-12:
iPhone
passcode
profile.

Figure 4-13:
iPhone VPN
profile.

To educate the end user about mobile device policies in the enterprise, you can use the following policy guidelines, which are directed toward your end users:

✔ You're required to use the recommended passcode policies on the device for basic device access security. You need to specify passcode length, duration, and patterns.

✔ You need to adhere to the VPN configurations, as defined in the enterprise mobile device configuration guide *[your published guide],* in order to gain access to the enterprise network.

  [If encryption strength (64 bit, 128 bit, and so on) is well-defined, you are protected against substandard VPN implementations.]

✔ You are required to set up the mail access configurations, as defined in the enterprise mobile device configuration guide *[your published guide],* in order to connect to the enterprise mail server to access your corporate e-mail.

✔ You are required to adhere to the web access policies, as defined in the enterprise mobile device configuration guide *[your published guide],* in order to gain access to the web.

The following user guideline applies to enterprise-issued mobile devices only; this cannot be mandated on employee-owned mobile devices:

> You are hereby being made aware that certain functions of the mobile device may be restricted when the device is connected to the enterprise network. These functions may include using the camera, installing custom applications, capturing screenshots, using external storage, and so on. Any subversions to these mandated policies could result in rescinding your rights to connect to the enterprise network.

Decommissioning policies

This section applies only to enterprise-issued mobile devices.

You may be called upon to decommission mobile devices for one of these two reasons:

- ✔ Accidental loss or theft of the device
- ✔ Willful violation of mobile device policies

In either of the preceding cases, the steps that need to be taken are similar. Therefore, the policies that you define for decommissioning mobile devices should also be consistent.

You can use the following guidelines to educate end users about decommissioning policies in the enterprise:

- ✔ You are expected to inform the IT department immediately upon loss or theft of your mobile device.

- ✔ Your device will be located and locked out, and all data will be erased as soon as possible.

- ✔ Any data loss as a result of the wipeout of the mobile device is your responsibility. IT does periodic backups; however, you are expected to follow the backup policies as well, especially if your device contains personal content, such as photos, music, and videos, for which the IT department bears no responsibility. Having a backup would allow you to quickly restore the configuration on a replacement phone.

- ✔ If the decommissioning is a result of policy violations, a replacement phone will not be provided to you. Furthermore, if you owned the device and violated policy, access to the enterprise network and its resources will be prohibited for up to a year. *[Change this based on your leniency threshold.]*

BlackBerry in particular has had the remote-wipe feature for years, and some of the newer mobile devices have this capability. iPhone owners can remote-wipe their device through the subscription-based MobileMe services. Additionally, Apple now provides the Find My iPhone app (shown in Figure 4-14) free of charge. It can run on any iOS 4.2 device (or newer), including iPhone 4, iPad, and the fourth-generation iPod touch. In the event of a loss or theft (or maybe just for the heck of it), individual users now have the power to remotely locate and wipe out their devices. While these capabilities certainly sound handy, it does not take away your responsibility toward protecting the enterprise assets.

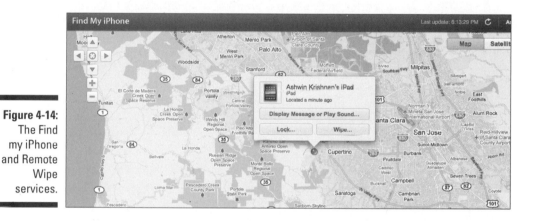

Figure 4-14:
The Find my iPhone and Remote Wipe services.

Creating Effective Monitoring Policies

Monitoring policies have for the longest time been veiled under shrouds of secrecy so that employees are not quite aware of how, why, and when their actions are being monitored, and the IT departments — read *you* — have also not been explicit and forthcoming about exactly what is being monitored and what policies govern the use of the data that is captured. It's time that we expose both the policies and governance models for the monitoring that you undertake so that everyone is keenly aware of them.

Unlike the device-based policies, which clearly distinguish between policies for employee-owned mobile devices and enterprise-issued mobile devices, monitoring policies are uniformly applicable toward all mobile devices regardless of their origin. The reason for this is once the mobile device connects to the network, it is incumbent upon you to be able to guarantee the security and integrity of the enterprise network and its assets. Therefore you need to do the monitoring agnostic of the type of mobile device.

Obviously, you have additional tools at your disposal for monitoring enterprise-issued mobile devices because you have local agents on the mobile device itself that you can exploit to gather this information. However, in the case of employee-owned mobile devices, you need to rely on the network exclusively to provide for monitoring capabilities.

To further muddy the waters, unscrupulous applications that employees download sometimes surreptitiously monitor detailed mobile device activity and sell it to advertisers and other scavengers. Once this is exposed, employees are extremely wary of any such monitoring apps, and you need to be all the more transparent in order to comply with your enterprise policies. In fact, a research project called TaintDroid (developed by researchers at Duke University and Penn State University) gives more power to advanced Android users by allowing this application to run in the background and alert users if any applications on their mobile device are shipping off their private information to a remote location, as shown in Figure 4-15.

Figure 4-15:
The
TaintDroid
application.

It is expected that more commercial tools will be developed that will allow users to take back control of their mobile devices, or at the very least, be made aware of any applications that are spying. Your job is to spy on your employees with a goal of keeping them compliant, so in effect, you should be doing what projects such as TaintDroid are doing on the mobile device: keeping tabs on everything that applications are doing and intervening where necessary.

You can use the following guide to educate your end users about the monitoring policies:

✔ All your activity when connected to the enterprise network will be monitored. This includes all enterprise applications as well as personal applications.

✔ Any data collected during the monitoring may be archived.

✔ Any willful obstruction of such monitoring may result in revocation of connectivity rights to the enterprise network.

✔ Purposely obfuscating data with the express intent of bypassing such monitoring is expressly prohibited.

The following guideline applies to enterprise-issued mobile devices only; this cannot be mandated on employee-owned mobile devices:

Even when the mobile device is *offline* (not connected to the network), it may still be monitored.

Because Android is an open environment where applications can be developed and marketed with little or no oversight by Google (as compared to the iTunes applications that Apple oversees closely), it is to be expected that a more fertile ecosystem of applications will thrive; some of them will give the user great control and visibility, the TaintDroid project being one example. In other less-open environments, like the iPhone or the Blackberry, it is far less likely that you can find monitoring applications that provide this level of scrutiny. However, where there is demand — in the form of users who are willing to pay for this level of scrutiny and take back control of their devices from opaque applications — there will be supply, so keep your eyes and ears open as these tools become available.

Protecting Devices with Application Policies

Application policies outline what applications users are permitted to use while accessing the enterprise network. Application policies are particularly crucial because the plethora of applications that users are able to download from mobile device vendors' websites, carrier application portals, and independent third parties is multiplying exponentially, as you can see in the chart shown in Figure 4-16. Your users are likely to innocently download a malicious application that causes havoc both to the user and, more to the point, the enterprise network.

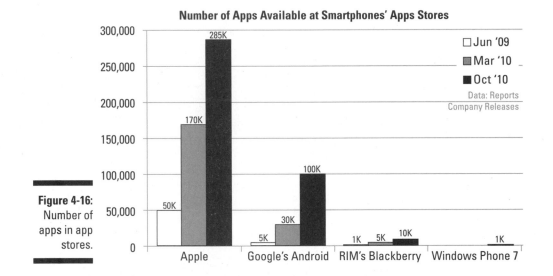

Figure 4-16:
Number of apps in app stores.

Typically, application policies can be categorized in the following subdomains, as shown in Figure 4-17:

- ✔ **White listing of approved applications that can be used by end users:** As with the other policies we discuss in this chapter, this policy can be interpreted as draconian. Clearly for enterprise-issued mobile devices, this policy can be justified to a degree because the mobile devices are enterprise assets and, therefore, the applications that reside on them can be controlled. However, even in this situation, there is a bonding that develops between the users and their mobile devices, and the users become *de facto* owners of these devices and assume moral authority over their usage.

  Quite obviously, for mobile devices that users own and bring into the enterprise, the actual application policies cannot be readily enforced on the devices. In this situation, you need to rely on the monitoring policies to ensure that enterprise compliance is being maintained when the device is connected to the enterprise network.

- ✔ **Profile settings for approved applications:** This policy applies to enterprise-issued mobile devices only, because it controls how applications can be configured and used. We discuss at length in the "Profile setting policies" section, earlier in this chapter, how to supply users with guidelines and recommended profile settings for applications. This is critical not only to ensure that the employee has connectivity and can actually use critical enterprise applications on the mobile device but also to limit the exposure that these applications have from a security standpoint.

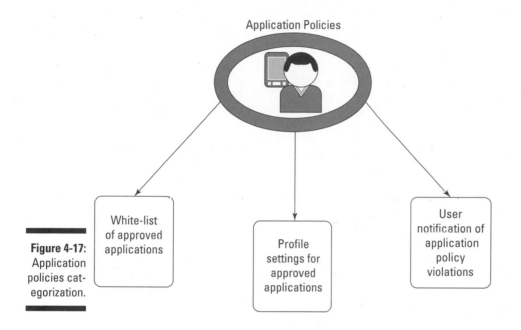

Application Policies

White-list of approved applications

Profile settings for approved applications

User notification of application policy violations

Figure 4-17: Application policies categorization.

For employee-owned devices, it's the user's prerogative what types of applications that she chooses to install, so there is no control you can exercise over that. However, you can impose restrictions on applications that connect to the corporate network like e-mail.

For instance, a very common profile setting for e-mail applications is to restrict download of attachments. This setting limits the exposure that the loss or theft of the mobile device can have because there aren't any residual documents on the device that an unscrupulous user can exploit. A less-restrictive version of this policy is to allow download of attachments to onboard flash only and not to removable media.

The BlackBerry Enterprise Server provides an arsenal of enterprise application settings that you can tweak to provide a custom environment that you are satisfied with. It provides a default application control policy that blocks third-party applications from running on a BlackBerry. Or, for the more liberal minded, it allows for application policies white lists to allow a controlled set of third-party applications to run on the mobile device. It can also get very precise and dictate data sets that the application is or isn't allowed to use, such as contacts, locations, photos, and so on.

✔ **User notification of application policy violations:** This policy applies to all mobile devices, enterprise-issued as well as employee-owned devices. The detection of these policy violations varies for these two categories. For the former, you may have additional tools at your disposal, including monitoring capabilities on the device in addition to network-based detection. For the latter, you need to rely on your monitoring capabilities on the network to detect out-of-policy application usage.

The response to application policy violations should be enforced in two steps. The first one is a warning notification indicating that the user is in violation of enterprise application policies. This notification could be in the form of a web page in the case of http-based applications, or e-mail in the case of other applications. After a certain number of warnings — spell this out clearly in your enterprise usage policy — the enforcement (the second step) needs to be more brutal. As with other policies, the remediation at this point is to inform the user that his right to connect to the enterprise network may be curtailed. Show users who is the master and commander!

Keep in mind that you can no longer rely on the application store vendors to be gatekeepers to the types of applications that appear on their store. In fact, the Android Market has been plagued with malicious applications. Apple keeps a much tighter leash on the applications that are allowed to be showcased on the App Store, but in spite of these stricter controls, Apple has had its share of unscrupulous apps show up as well.

At the highly regarded Black Hat DC 2010 conference (a premier security event to discuss security breaches and vulnerabilities in hardware and software), software engineer Nicolas Seriot disclosed that two iPhone applications had been pulled out from the App Store because of privacy violations:

✔ Aurora Feint (in July 2008) harvested all the contacts from a user's iPhone and uploaded the entire contents unencrypted to the game developer's server.

✔ MogoRoad (in September 2009) uploaded users' phone numbers to the developer's server, and the users were then harassed by live telemarketing agents who were peddling products. These are just two applications but are representative of what apps on the loose can do.

Storm8, another popular game developer for the iPhone, was charged in a complaint field at the U.S. district court in Northern California in 2009, about its games surreptitiously collecting user information, mobile numbers in particular.

As we discuss in Chapter 2, jailbroken iPhones should be a source of concern for you. When an iPhone is jailbroken, there is no gatekeeper to ensure that the applications running on it aren't well-packaged malware. Until December 2010, Apple had a jailbreak-detection API that applications could use to detect whether an iPhone was jailbroken, but for some reason, Apple decided to pull support for this API. You still have a number of third-party applications that you can use to perform similar detection functions should you choose to limit the enterprise exposure to jailbroken iPhones by detecting them and branding them as unsupported devices, primarily because of the opacity surrounding the applications that are actually running on those devices.

To make matters more interesting, some good citizens are actually using the jailbreaking of iPhones to make attacks more difficult. Case in point is a tool called Antid0te, created by Stefan Esser of SektionEins, which adds Address Space Layout Randomization (ASLR) onto jailbroken iOS devices. What this tool does is randomize key memory locations to make it more difficult for certain attacks to locate their target data.

More such applications are likely to emerge, but be wary of embracing apps that don't have much pedigree in the form of a big reputable firm behind them. You should avoid recommending any applications as part of your defense arsenal that are made by unknown parties. Although some of them may sound promising and make bold claims, these could be cloaked malware or poorly written applications that do more damage than remedy.

Regardless of these advancements, your defensive posture should be taking out support for devices that are not trusted (also known as those that are no longer supported by the manufacturers because of active intervention by the users that change the posture of these devices).

Your users live in dangerous waters regardless of whether they are aware of it, and it falls upon you to protect them from the predators in the guise of applications that threaten to launch an attack anywhere and anytime.

Case Study: AcmeGizmo Mobile Device Security Policy

In this section, we revisit our ongoing case study of AcmeGizmo, which we introduce in Chapter 1. Ivan, our IT administrator, has been doing a lot of research into possible solutions for his mobile device problem, and has also learned quite a bit about the possible issues and threats that he needs to concern himself with when providing enterprise access to mobile devices. He has come up with a fairly comprehensive security policy that covers not only the types of technologies he is going to deploy but also acceptable-use guidelines for employees of AcmeGizmo to follow when using their mobile devices to access corporate data.

Ivan decided to implement a security policy that will allow for maximum productivity but is restrictive enough to meet AcmeGizmo's overall corporate security needs. The majority of the policies in the example security policy for AcmeGizmo are covered in the case study sections in other chapters, but a couple are covered here.

- ✔ Ivan decided that all employee devices must be registered and approved both by the corporate help desk and the employee's manager prior to accessing the network. This will not only help Ivan track the devices connecting to the corporate network, but also ensure that the employee's manager feels that mobile device access will benefit the individual's productivity.

- ✔ Ivan decided to include a password policy. Many mobile devices available on the market do not force the user to have a password, but because these devices will be able to access sensitive corporate data, Ivan wants to ensure that a lost or stolen device is adequately protected from someone picking it up and easily having access to AcmeGizmo data. For this reason, Ivan wants to ensure that all devices have a lock time-out of 10 minutes, meaning that if the user doesn't use the device, it will automatically lock within 10 minutes of inactivity.

  In addition, all lock passwords must be at least six characters, with a mix of characters, numbers, and symbols. This is in addition to the one-time password solution being used to connect AcmeGizmo devices to the network via SSL VPN.

- ✔ Many mobile devices now ship with built-in full disk encryption, but many more do not. Ivan has decided that all devices that ship with full disk encryption must have it enabled (if it's possible to disable it). In addition, he has chosen third-party encryption software that will be installed on all devices that do not have built-in encryption.

- ✔ Finally, Ivan established a couple of requirements that ensure that employees follow reasonable acceptable use policies and procedures. These requirements ask employees to use care with their mobile devices to ensure that they do not carelessly leave them unattended, as an example. Additionally, he requests that employees do not download and save sensitive data to mobile devices, as an extra precaution, just in case the device does get stolen. Finally, he outlines a process of contacting the AcmeGizmo help desk immediately if an employee misplaces a device or if the device is stolen.

Here's the security policy that Ivan wrote for distribution to all employees. As with all good security policies, special care has been taken to ensure that the policy is written in language that end users can understand. Ivan also attempted to keep it as brief as possible, which helps to minimize the confusion that too many rules and regulations can cause.

Overview: This is the AcmeGizmo policy on mobile device usage. A *mobile device* is any PDA, smartphone, or other portable device that has the ability to access the corporate network and can store sensitive corporate data. (Note that laptops are not included in the definition of a mobile device.)

Purpose: This policy has been designed to ensure that AcmeGizmo employees protect mobile devices and corporate data from theft and loss.

Policy:

✔ All devices used by AcmeGizmo employees to access *any* corporate data must be approved by IT security and the employee's manager prior to use.

✔ All devices must run one of the following operating systems:

- Apple iOS 4.0 or newer
- Google Android 2.1 or newer
- RIM BlackBerry (any version)
- Windows Mobile 6.5 or Windows Phone 7
- Nokia Symbian

No other mobile devices are permitted to access the AcmeGizmo corporate network.

✔ All devices must have a lock time-out of 10 minutes or less, as well as a lock password with a minimum of six characters. All passwords must contain at least one nonalphanumeric character.

✔ All devices must have the Junos Pulse client installed and running at all times. This software includes a virtual private network (VPN) component that securely connects the device to AcmeGizmo. It also includes antivirus and personal firewall software. In addition, it gives AcmeGizmo IT the ability to remotely lock, wipe, and locate any mobile device if it is lost or stolen. End users connecting devices to the corporate network agree to installation of such software on their mobile devices.

✔ All devices must have encryption enabled, or a third-party full disk encryption product installed.

✔ Employees must employ all reasonable means to ensure that mobile devices are not lost or stolen, which requires extra diligence on the part of every employee to mitigate carelessness.

✔ Sensitive corporate data must not be downloaded to mobile devices. Data access is restricted to corporate e-mail, the corporate CRM application, and the company intranet. Sensitive attachments received via e-mail must not be stored permanently on the device disk drive or on any removable media.

✔ Employees must immediately report any lost or stolen devices to the help desk. Lost or stolen devices may be subject to a *full wipe* (erasure of the disk) when reported lost or stolen. This may result in the loss of not only corporate information but any personal employee information stored on the phone as well.

✔ Any requests for exception to this policy must be approved by AcmeGizmo's IT security.

Chapter 5

Managing and Controlling Devices

· ·

· ·

*T*he smartphone explosion has been fast and furious. Almost overnight, tens of millions of smartphone devices have been sold. Because you're reading this book, we assume that some of them are in the hands of your users, and they are demanding access to corporate data from these devices.

To compound the problem, many of these devices are probably owned by the end users themselves, rather than by your organization. The personally liable (user-owned) device introduces a whole new set of concerns when you're trying to manage and control these devices. Typically referred to as the "consumerization of IT," today's users demand access from their device of choice. Gone are the days when the IT department could provision a single type of computer and a single mobile device to all users that required them. Sure, some strongholds still exist, but with each passing day, more departments are bending to the pressure and beginning to allow these types of devices to access corporate data.

Whether you're provisioning and managing corporate-liable (company-issued) mobile devices or allowing end users to purchase their own devices, or both, this chapter helps you get a handle on what solutions are available to ensure that the mobile devices in your enterprise have the appropriate policies that will allow you to feel comfortable with even the most basic aspects of mobile device security, such as controlling whether a device has a password. Typically referred to as *mobile device management* (MDM), these types of solutions help you provision and maintain policies on your organization's mobile devices remotely.

This chapter also covers application provisioning. When you allow mobile device applications into the network, you need to provide the necessary tools to ensure that your end users are productive. This might include controlling which applications can or can't be installed on mobile devices, but it might also include an emerging set of technologies that allow you to create your own enterprise app store for your end users.

Finally, the chapter closes with a discussion on our ongoing case study, AcmeGizmo, and what mobile device management and control attributes were most important in the fictional company's smartphone deployment.

Managing Your Mobile Devices

Mobile device management (MDM) is a broad and ever-expanding category of product offerings from several vendors that allows an organization to manage the full lifecycle of its smartphone deployment — from initial configuration, including security policy configuration, to support and troubleshooting and reporting. These solutions provide much-needed central management and provisioning applications for your enterprise to ensure that the mobile devices connecting to your network are doing so only when they have been properly configured and secured.

Because the mobile device management space is still relatively new and because it is a quickly growing market, there is a lot of overlap across various types of solutions, all of which claim to be MDM solutions. In some cases, vendors that have traditionally been in the telecom and expense management category have added a large number of MDM functionalities. In other cases, security vendors have felt the need to add MDM capabilities to their products. And of course, there are a number of companies that specialize exclusively in MDM. For this reason, choosing an MDM vendor can be easy or difficult, depending on your role in the organization and your primary goals.

The primary focus of this chapter is on the elements of MDM that have a direct impact on the security of your mobile device deployment. That said, there are a few nonsecurity areas of MDM that have an impact on your organization's ability to deploy and maintain security on these devices; we also cover those areas. In addition, we cover the two most popular protocols leveraged by MDM vendors: Exchange ActiveSync (EAS) and Open Mobile Alliance Device Management (OMA DM).

Managing devices over the air

One of the most important elements of any mobile device management product is the ability to manage devices over the air (OTA). Given the mobile nature of these devices, no MDM strategy would be successful if it required devices to be physically connected to a machine or network periodically.

OTA management is available for every manageable function that we talk about in this chapter. Not only does this allow you to manage devices without ever physically touching them, but it also allows you to centrally manage groups of devices all at once, greatly reducing the time spent on this task.

Commands are sent to mobile devices one of two ways:

- ✔ **SMS:** One popular way to send commands to a mobile device is to use short message service (SMS — text messages). For example, an SMS might be sent to a mobile device as part of an enrollment process, with a URL pointing to the MDM server, allowing the user to add a new device under management.

  One of the most compelling attributes of SMS is the fact that it is available nearly everywhere. A common example is when a device is roaming into a different country — the user might choose to turn off data services, rendering push notifications useless, but SMS will likely still be available.

  A downside of SMS management is that it is available only on devices that are connected to a mobile network with a 3G or 4G radio. Increasingly popular Wi-Fi–only tablets and other mobile devices cannot be managed via SMS.

- ✔ **Push notification:** Push notification services are available on many popular mobile device platforms. *Push notification* accomplishes the same goal as SMS, but leverages Internet-based communication channels to manage the device.

  Push notification mechanisms are attractive because they leverage a reliable communications mechanism so that the sender of the notification knows whether that notification reached the end destination. Additionally, a push notification can handle any amount of data and is not generally subject to per-message charges, as is the case with SMS.

  A downside of using an Internet-based management channel is that it requires Internet access in order to work. If a user has disabled her data connection, such as when roaming, important updates and notifications might not be possible on that device until she connects to the data service once again.

Your MDM solution might leverage one or both of these techniques in order to manage mobile devices. Both have some weaknesses, as described here. The most robust MDM solutions will leverage both of these techniques as applicable for specific devices and situations.

Initial provisioning workflow

It is useful to understand how the processes of initial provisioning and ongoing management of a smartphone device work. In this section, we step through the initial provisioning process.

Note that the processes described here are the most common processes. In some cases, they have been simplified to only the portions of the processes themselves that are directly relevant to understanding how the device goes from an unknown device to a fully provisioned and manageable device.

Figure 5-1 shows the typical flow of initial provisioning of a device as described here. After you decide to manage a device with an MDM application, follow these steps:

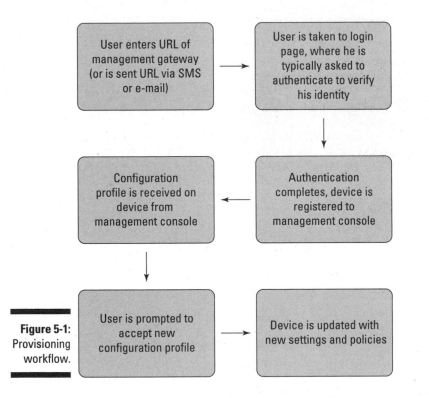

Figure 5-1:
Provisioning
workflow.

1. **Connect the device to the management console or server so that it can be provisioned.**

   This means that the URL of the management server needs to be entered on the device. You can use one of these two methods to do that:

   • *Tell the user the URL or send an e-mail with a clickable URL to the user.*

   • *Send the URL via SMS from the management server to the device.*

   Some solutions initially require the user to download a client application from the smartphone application marketplace. With these applications, the flow is similar, but rather than entering the URL into her mobile phone browser, the user enters it into the management client application.

2. **Have the user click the URL or enter it in the appropriate application and authenticate at that login page, verifying the user's identity.**

   Upon completion of authentication, the management console typically completes a registration process. At this point, it might verify which policies and configuration to apply to the device based on who the user is, her role in the organization, and the type of device that she is registering.

   The management console then sends the appropriate configuration information to the mobile device. Often, this is in the form of a configuration file that the device knows how to interpret.

3. **Have the user accept and install the new configuration file when prompted.**

   This step does involve end user input, and it is important from a security perspective. You certainly don't want your users accepting updated configuration files from rogue management servers. Make sure that you train your end users on the dangers of accepting unknown configuration files or connecting to other management servers.

   After the user has accepted the new configuration file, the device installs the file and is updated with its new settings and policies (the settings and policies described throughout the remainder of this chapter).

The process described here is the typical over-the-air provisioning process employed by mobile device management solutions. There are other options for deployment of the configuration file to the mobile device, including

- ✔ **E-mail:** The configuration file can be e-mailed to the recipient as an attachment, which is then installed on the endpoint device. A word of caution here: If you have enforced a policy that restricts the downloading of attachments from e-mail, this would actually conflict with that policy. Ensure that you are not so restrictive that you prohibit your users from installing the security software that you want to see on their devices!

- ✔ **Web-based delivery:** The file is placed on a web server that the user browses to in order to install the configuration.

- ✔ **Direct connection to a PC:** This option leverages a configuration utility, such as the iPhone Configuration Utility. In this case, the user physically connects his device to a desktop or laptop via USB, and the config file is installed on the device as the device synchronizes to the software on the PC.

Ongoing management

Initial provisioning and configuration of a device is only the first step. After you have devices under management, you are likely to want to make changes to policies and configuration over time.

After the device has gone through the initial provisioning workflow, as discussed earlier in this chapter, it is subject to ongoing management by the MDM server. As you make updates and changes to the configuration, the device will accept those updates with minimal additional end user interaction.

Figure 5-2 describes the ongoing management process in more detail. In the following list, we take a closer look at each step:

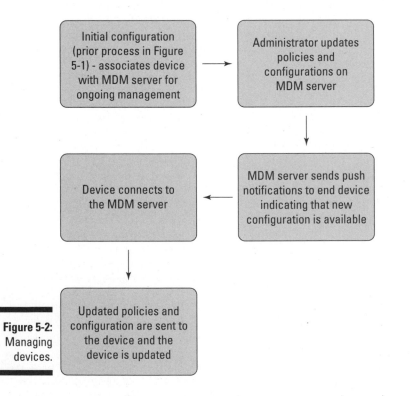

Figure 5-2: Managing devices.

1. **Complete the initial provisioning process as described in the preceding section.**

    After this is completed, a relationship between the device and the MDM server is formed, providing the basis and channel for ongoing updates.

2. **Make any policy and configuration changes that you would like on the MDM server.**

This allows the MDM server to generate an updated configuration file. The MDM server then sends a push notification to the device, indicating that new configuration information is available.

The device then connects to the MDM server, and the updated policies and configuration are sent to the device, typically via a configuration file. The device then executes the new configuration.

Configuring security policies

Several chapters in this book deal with protection mechanisms for mobile devices, such as anti-malware software for the device, or the ability to remotely wipe the data from the device if it is lost or stolen. Those mechanisms are an extremely important part of any smartphone security deployment, but before you begin to roll those out, there are a number of basic configurations or policies that you should set on the device to ensure that it has an appropriately secure configuration. These configuration policies include setting an appropriate password, encrypting data stored on the device, setting network configuration such as Virtual Private Networks (VPNs) and Wi-Fi, and restricting other features of the device itself.

Implementing password policies

Just about every smartphone on the market includes the ability to set a password that is required to access any of the device's features (other than the emergency call function, which is required by law to be available regardless of whether the device is locked). Unfortunately, the same devices that allow this functionality almost always have password policies turned off by default.

The lack of a required password is a huge problem. Without one, any person who picks up the device can fully access everything on the device, including the data and applications. This is a huge security exposure, and you must be absolutely certain that everyone who accesses your corporate data with a smartphone has a password set on the device — hopefully one that meets your corporate password policy guidelines.

The onus is on you to ensure that every device connecting to your corporate network is configured to support your policies. There are a number of password policy settings supported by most devices. Most of these policy settings are best practices for any password in corporate or personal environments, regardless of what type of device or application you are trying to secure.

Here's a rundown of password policy settings:

- **Password required:** This is the first and most obvious setting. It would be great if all smartphone manufacturers required an unlock password on all of their devices — not only for enterprises but for end consumers as well.

- **Minimum password length:** This one is simple. The longer the password, the more difficult it is to crack. Password length settings are a double-edged sword. If you make them too short, they are easy to overcome; if you make them too long, they become difficult for users to remember — and users can end up locked out of their devices.

  Industry best practices typically recommend a minimum password length of 6–8 characters.

- **Password complexity:** At its simplest, this means that any password must contain a mix of several different types of characters. For example, an organization's password policy might dictate that a password must contain a mix of upper- and lowercase letters and at least one number, symbol, or punctuation mark.

  The more of these types of special characters you require, the more difficult the password is to crack, but you also run a higher risk that the user will forget the password and get locked out, or write it down where others can find it. Both of these are poor outcomes that you want to avoid.

  Every smartphone on the market allows the user to answer the phone without typing in a complicated password. Additionally, emergency phone calls are always allowed from these devices without requiring the user to enter the password. All other outbound calls, as well as access to any other features of the device, require the use of the password.

- **Password aging:** Your password policy for your mobile devices should have a password-aging component. This is the length of time between forced changes of a user's password. For example, you might specify that users change their password once per quarter or twice per year. This component of a password policy is important because over long periods of time, the likelihood that someone will find out or guess a user's password increases.

- **Password history:** This setting allows you to control the number of new passwords that must be used before a user can begin to reuse prior passwords. Your password-aging policy won't be nearly as powerful if the user switches every quarter between the same two passwords. Most good password policies require at least four unique passwords before an end user can reuse a prior password. Device settings typically allow for much higher numbers than that, enabling you to effectively ban reuse altogether if you so choose.

✔ **Idle timeout:** This setting allow you to specify the amount of time that a device can remain idle, with no user input, before it is locked automatically. The idle timeout setting is an important part of your password policy, because without it you must rely on end users to always lock the devices themselves, which is a very unreliable form of security. With this component of your policy, you want to strike a balance between security and usability. If your devices automatically lock after 30 seconds of inactivity, for example, your users will be constantly re-entering their passwords and losing productivity. If, on the other hand, the devices don't lock for 24 hours after the last activity, you have a huge window of exposure if that device is lost or stolen. A best practice recommendation is to set devices to lock after 5 minutes or fewer of inactivity.

✔ **Maximum number of incorrect passwords:** If someone steals a mobile device that has been locked, they are likely to try getting into that device using brute-force password guessing. For this scenario, you want to have a policy enabled that specifies how many times an incorrect password can be used before the data on the device is deleted. Without this type of policy, the thief can continue to guess passwords indefinitely. A best practice for this policy is to set a maximum of 10 incorrect passwords before the device is remotely wiped.

Make sure that your end users know that if they type in the incorrect password too many times, their data will be deleted. This is a good thing to inform users of both in security policies that they review, as well as in any training that you conduct to educate them in the proper use of their mobile devices.

Of course, it's also a good idea to ensure that you are continually backing up data on the device, just in case it ends up getting wiped. You want to protect your corporate data, not lose it altogether. Chapter 12 covers how to back up and restore data onto existing or replacement devices.

Beyond settings that you can control via configuration on the devices themselves, any good password policy also has an end-user training and education component. The same must hold true for your mobile device password policy. You should train your end users to follow these guidelines:

✔ **When creating passwords, never use words found in a dictionary or commonly used words.**

For example, an end user should never use his child's name or the name of his pet as his password.

✔ **Never write down your password —** *anywhere.*

This means, of course, that end users shouldn't write down their passwords on a sticky note on their desk or monitor, but it also means that users shouldn't write down passwords in locations that they feel are more secure either — such as in their home or on a piece of paper in their purse or wallet.

✔ **Never tell *anyone* your password.**

This is true of the user's boss, the IT administrator, friends, family, and so on.

✔ **Never talk about the type of password that you use or the password format.**

If a malicious person knows that the password contains a number, is six letters, and is based on your most recent vacation, that person has a strong start on cracking your password. Ensure that your end users don't make it easy for someone to beat your password policies.

A good combination of strong password-enforcement policies along with end user education and training will ensure that your risk exposure, should a device be lost or stolen, is minimal. This definitely does not mean that a strong password policy is all that you need; this is just one small component of an over-arching security strategy that you will put in place based on this chapter and on the topics covered in the rest of the book.

Removing prohibited applications

Increasingly, MDM solutions allow an organization to restrict the user's ability to download or use prohibited applications. In some cases, the MDM solution has the ability to uninstall a prohibited application. This is typically in addition to any application removal that an antivirus solution would provide for the device.

Antivirus typically handles applications that are known to have malicious intent. The ability to remove applications via an MDM solution is geared more toward applications that, for one reason or another, the organization does not want to have installed on the device. The reasons can range from security concerns about the application itself (such as the potential for the application to inadvertently leak sensitive information) to productivity concerns (if a user is spending too much time using their social networking application, for example).

Encrypting data

Encryption of *data at rest,* data that has been downloaded to and will be stored on the smartphone itself, is an important policy to set. Encrypting data prohibits someone from connecting a stolen smartphone to a PC and synchronizing sensitive data from the device to her PC, as an example.

Depending on the operating system platform and device, encryption functionality may or may not be built into the mobile device. In some cases, such as with Apple iPhones running iOS 4, encryption is built into the device. In other cases, encryption is not provided in the base device and operating system, so third-party software is required to accomplish encryption. Increasingly, operating system vendors are including encryption capabilities in the operating system itself, so the primary task is to ensure that encryption is enabled.

For those platforms that include encryption, the task of managing that encryption should be handled by your mobile device management solution. You want to ensure that encryption is enabled across the entire device, especially for any data downloaded to the device, including files, application data, and so on.

Be very careful to ensure that when you enable encryption, you know exactly what is being encrypted and what isn't. For example, the default encryption policy on a device might encrypt data on the device disk itself — e-mail, contacts, calendar, and personal documents — but might not encrypt data saved to removable media such as an SD card, for example. If that's the case and you simply implement default encryption settings, you will be leaving yourself open to a critical security vulnerability and an easy way for data to be lost or stolen.

Configuring network settings: VPN, APN, and Wi-Fi

Another critical security related feature with most MDM solutions gives you control over network connections and the way that each mobile device connects to the corporate environment. Here are several ways that mobile devices connect to networks:

- **Virtual private network (VPN) access:** Most MDM functions allow you to configure VPN access requirements on a mobile device. It is critical that all connections to your enterprise environment do so through a VPN, and setting the policies on behalf of the user makes it that much easier for the user to be in compliance with your policies. Typically, the MDM solution allows you to specify a VPN gateway, along with the type of user authentication required and any other information required to securely connect to the corporate network. Chapter 7 discusses VPNs in detail.

- **Private access point name (private APN):** Private APNs are much like VPNs from a security perspective. The APN configuration specifies the point where a mobile device can access an IP network. Many service providers globally provide a private APN service to large customers, enabling them to separate their data traffic from that of other customers, or from customers on their public APNs. MDM solutions can sometimes configure mobile devices to support private APNs.

Private APNs are a proxy for an IPsec or SSL VPN only if the device is connected directly to the carrier network. For devices that do not have 3G or 4G service, or when devices are connected to Wi-Fi but not to the carrier network, the level of data security and segmentation that a private APN provides is no longer available. Make sure that you are aware of this and plan for it in your security and Wi-Fi access policies.

- **Wi-Fi access:** Many mobile devices on the market today have multiple radios — one for accessing the 3G or 4G network and another for Wi-Fi access. MDM solutions allow you to configure mobile devices to seamlessly connect to Wi-Fi access points of your choosing.

For example, you may have an enterprise wireless local area network (WLAN) deployment that you want devices to access when the user is on the corporate campus. MDM allows you to specify the service set identifier (SSID) of the WLAN, and also specify the security settings, such as encryption type and password, required to have the device seamlessly join the network when it's within range. It is always a best practice to enable encryption on your wireless LAN.

Many MDM solutions also allow you to select whether Wi-Fi access is allowed from a mobile device at all. This truly prevents users from connecting to insecure or untrusted wireless networks; however, it also has the likely impact of limiting the users' productivity or forcing them to use more expensive 3G data services, even if free or low-cost Wi-Fi is available in their current location. As with other security mechanisms, disabling Wi-Fi is not without tradeoffs in end-user experience.

Restricting device functionality

There is a wide range of device functionality that you may want to restrict or control on your organization's mobile devices in the interest of security. Not all of the following features are available in all MDM platforms or on all smartphone operating systems:

- ✔ **Application store access:** On today's smartphones, application stores are extremely popular, with some form of an app store on every major smartphone platform. These stores or marketplaces have revolutionized the way that applications are delivered to a device, and they make it as simple as a single click to download and install an application. These applications are typically free or very low cost, drastically reducing barriers to new software acquisition for end users.

  The primary issue with app stores, however, is that they make it much easier for end users to inadvertently download and install malicious software such as a virus or spyware, as described in the later section "Pros and cons of consumer app stores." The amount of security scrutiny that posted applications get varies greatly depending on the application store in question.

  It probably does not make sense to restrict access to application stores because applications are a big reason why smartphones have become so popular. Your end users will likely revolt if you try to restrict access. A more feasible option may be to allow access, but to protect users' devices with many of the other security best practices described in this book.

- ✔ **Third-party application downloads:** This restriction applies to applications added to the device outside of application stores. Because most smartphone operating systems include the ability to install software outside of the application store, this restriction provides additional protection against unwanted applications installed through other mechanisms, such as desktop synchronization. This policy is important because outside of the sanctioned application stores, your users don't really know who has created the software that they have downloaded. Malicious entities look for these types of application stores precisely because they know nobody is forcing them to validate their identity or reviewing the application before it is posted.

✔ **Removable media access:** This setting controls whether an end user can copy data and files to removable media such as an SD card. Many organizations underestimate the threats of data leakage via removable media, but it is a very real threat. It is all too easy for users to copy data onto such media and then lose the media itself because it is typically the size of a postage stamp. It is a good idea to restrict what your users can move from their mobile device to removable media.

✔ **Screen capture:** This type of policy controls the user's ability to take screenshots of what is on the device screen and make that data available to applications on the device. Allowing screen captures is a security vulnerability because a user can essentially remove data from the device without physically e-mailing or sending an attachment, for example. Ensure that if you have very sensitive data on the device, you protect against all mechanisms of data leakage, including screenshots.

✔ **Clipboard operations:** Similar to screen capture, these functions allow you to control whether an end user can cut, copy, and paste text on an end device. As with application stores, these features are an important part of your users' overall smartphone experience, so you likely won't be successful in restricting these capabilities even if you want to, except in the most sensitive environments. Properly training your end users not to inadvertently leak data is the more likely approach that you will be able to take.

✔ **Bluetooth access:** In the past, Bluetooth was viewed as a potential mechanism for breaking into a device or distributing viruses. All smartphones today, however, include a security functionality that requires the use of authentication before pairing devices, greatly reducing the risk of a malicious person *bluejacking* a device within close proximity.

✔ **Use of the device's camera:** In certain situations, such as in defense-related organizations, users are not allowed to use the cameras on their mobile devices. Typically, an MDM solution gives the organization the capability to enable or disable use of the camera on mobile devices under management.

✔ **Access to consumer e-mail accounts such as Gmail or Yahoo! Mail:** Your users might make it difficult to enable this type of policy on devices that they also use for personal reasons, but for a corporate-owned and -issued device, restricting access to consumer e-mail accounts might make perfect sense. Microsoft introduced this policy in EAS (Exchange ActiveSync) 12.1, and it allows the administrator to control whether end users can access consumer e-mail accounts in addition to their corporate-provisioned Exchange e-mail accounts.

In this section, we have described a range of possible device restrictions that you have at your disposal — a very powerful proposition. Be careful, though, because that power can easily be abused. These restrictions have the potential to severely cut down on the functionality and usability of a mobile device. From a security perspective, that sounds great. From a productivity perspective, however, it's not very good. At the end of the day, your job is to enable users to be productive without putting sensitive corporate assets at risk; you need to balance restrictions with the needs of your users to get their jobs done.

Open Mobile Alliance Device Management

In many networking and security areas, open standards play a critical role in ensuring operability across different devices. Mobile devices are no different. The Open Mobile Alliance (OMA) was formed in 2002 as a consortium of vendors with a common goal of interoperability for new and emerging mobile technologies. Within OMA, the Device Management Working Group is responsible for management of both software and configuration on mobile devices.

According to the Device Management Working Group Charter, the working group is responsible for creating standards related to these items:

- ✔ Initial configuration of mobile devices
- ✔ Ongoing installation and updates of configuration
- ✔ Remote retrieval of management information from devices
- ✔ Processing of events and alarms generated by mobile devices

Devices that implement support for OMA Device Management (DM) may implement a subset of the various functions covered by the standards, so not all devices will support every feature. The OMA DM standards were designed with mobile devices in mind, ensuring that battery life, network throughput, and security are covered and adapted to the special needs and requirements of the typical mobile device.

As your organization proceeds with purchase decisions related to mobile device management, support for standards might be something to take into account. The most compelling reason to look at an MDM solution that supports OMA DM is that as new devices come onto the market, they are likely to support open standards like OMA DM, ensuring that you have investment protection from your MDM solution and the ability to rapidly adapt to new devices as your users attempt to bring them into your corporate network.

Exchange ActiveSync

Exchange ActiveSync (EAS) is a proprietary protocol developed by Microsoft that has been widely licensed and adopted by device operating system vendors, and has become a de facto standard over the past few years. Most smartphone operating systems sold today support ActiveSync. Microsoft developed EAS as a synchronization protocol for Microsoft Exchange, but it has been adapted over time to include more device security and management functionality.

The latest versions of EAS support a number of security policies and settings, some of which include

- ✔ Various options for setting password policies, including password length and complexity requirements.
- ✔ The ability to remotely lock or wipe a device.
- ✔ Tools to wipe and/or encrypt removable media, such as SD cards.
- ✔ Controls for whether a device that does not support all of the EAS password policies can connect to the Exchange Server.
- ✔ Password expiration and policy refresh intervals.
- ✔ Policies that control whether attachments can be downloaded to the endpoint device.
- ✔ The ability to disable Wi-Fi, infrared ports, Bluetooth, and cameras.

Many enterprises implement these EAS security policies when they first allow mobile devices to access their network. Typically, the mobile devices connect directly to the e-mail server via the Internet.

Always ensure that you are using HTTPS for devices connecting to the mail server via the Internet. You must never allow direct HTTP access because HTTP is unencrypted, and you should never have sensitive corporate data transiting the Internet unencrypted. Unencrypted data, if intercepted by malicious entities, can be captured and read by the receiver of that data. By encrypting e-mail via HTTPS, that data cannot be read, even if it gets into the wrong hands.

A question that might be on your mind is whether it is still necessary to deploy additional security mechanisms if the aforementioned EAS security policies have been put into place. The answer is a resounding yes. The EAS policies are only a small part of security best practices for deploying mobile devices. EAS covers some of the basics — password policies and the ability to remotely wipe or lock lost or stolen devices — but it doesn't cover the majority of the security best practices covered throughout this chapter and throughout this book.

For example, EAS allows all traffic to and from the Exchange Server to be encrypted, but it does nothing to protect traffic to and from other application servers that the user might want to access. Additionally, EAS provides no mechanism for controlling access to application stores or downloading of third-party applications, limiting your ability to control the applications that are downloaded onto your organization's devices.

Exchange ActiveSync provides some attractive security benefits, but it is far from a complete security solution. It might be part of a layered approach to smartphone and mobile device security, but it isn't built to stand alone and secure devices on all levels.

Controlling Applications

After a device is provisioned with the appropriate security policies and configurations, you need to deploy the appropriate applications to the device. In some cases, you are doing so in order to ensure that the device is fully secured. For example, you might deploy an anti-malware application to the device. In other cases, it is more about ease of deployment — there might be several enterprise applications that the users in your organization require, and you want to ensure that each of those applications is available.

The most popular way that consumers acquire applications for their smartphones is through application stores provided by their device manufacturer or by their service provider. This is a potentially useful distribution mechanism for the enterprise as well, but only for certain types of applications — and this approach is not without risks.

Pros and cons of consumer app stores

There are pros and cons of using consumer application stores in enterprise environments. On the plus side, this is one of the most ubiquitous distribution mechanisms available. Just about every smartphone on the market today has an immediately available application store, and users have become very comfortable with using of application stores as a way to download and install applications onto their smartphone device. There are, however, very clear dangers and drawbacks to taking such an approach.

One of the dangers is the fact that many consumer application stores are not closely monitored from a security perspective. Some application stores, such as Apple's App Store, have fairly stringent screening processes that are required before an application can be posted. Others, such as the Android Market, have less stringent screening processes, greatly increasing the likelihood that an application that makes its way into the application store is malicious. In either case, the risk is there.

Increasingly, bad actors are sneaking malicious applications into these app stores, and if you leave it to your end users to find software on their own, they might end up with software that simply looks like the software they were seeking. An application that appears to be a commercial customer relationship management (CRM) application might actually be spyware.

In 2010, several online banking apps were posted to the Android Market. Consumers were led to believe that these applications were published by several very large banking organizations. It turned out that these were not legitimate banking applications; they were posted by a third party that made them appear to be official online banking applications. In this case, the applications were designed for phishing — a clever way to steal end users' online banking login credentials. With hundreds of thousands of applications in these app stores, it is easy for a user to mistake a fake and malicious application for a legitimate one. You certainly don't want your corporate data to fall into the wrong hands this way.

Provisioning applications to mobile devices

It is not a good security practice to leverage consumer application stores to deliver applications to your users' mobile devices, so you need to opt for one of the following delivery options, the most popular of which is over-the-air provisioning.

Over-the-air application provisioning

Over-the-air application provisioning is exactly as it sounds: Many MDM platforms offer the ability to deliver applications to smartphones over the air, leveraging many of the same mechanisms described throughout this chapter for provisioning policies and configurations. With this method, you select the applications to provision and the devices to which those applications should be provisioned and deliver them wirelessly to those devices.

Web-based provisioning

MDM solutions provide an easy and packaged method for provisioning applications, but there are easier, less expensive options. One method is simple web-based provisioning, where your organization hosts the appropriate applications on a web server. The employee then clicks the appropriate URL to retrieve the installation package from the web server. The URL can also be provided to employees via SMS or e-mail to make it easier for employees to know where to get the application.

Applications catalog

Some solutions on the market offer you the ability to create a catalog of corporate-approved or -created applications from which an employee can choose. These applications can be delivered to the device in one of several ways, including direct download from an application store, over the air delivery, or direct installation of the application to the device through an SD Card or USB connection to a PC.

In some cases, this online application catalog is nothing more than a pointer to the URL for applications that are hosted on the application store of the relevant app store vendor. For example, you might provide a link to an off-the-shelf customer relationship management application. The advantage of taking this approach versus having the employee look for the appropriate application is that it cuts down on guesswork and virtually ensures that the employee selects the appropriate application, and not a malicious app masquerading as a legitimate application. This approach also makes it easier for employees to get everything that they need to be productive on a smartphone device; they can find all of their recommended applications in one location.

Some operating system vendors also allow for the creation of enterprise application stores, which are entirely separate from the consumer version of their application stores. For example, in 2010, Apple made it possible for enterprises to create their own application stores for iOS devices through the iOS Enterprise Program (part of Apple's iOS Developer Program). There are also some third-party and open source tools that allow you to create your own Google Android application stores.

With these enterprise application stores, not only is application distribution limited to only selected applications, but you can also create and distribute your own applications outside of the stringent restrictions sometimes put on applications before they can be posted to consumer application stores. Enterprise application stores also allow you to keep your applications from others, limiting distribution to employees only.

Blacklisting and removing applications

Along with provisioning of applications comes monitoring and control. As with PCs, many organizations have policies that dictate the types of software that can be present on a corporate-owned and -issued device, for example. In this case, the ability to inventory, blacklist, and remove applications becomes important.

The first step in the process is taking an inventory or snapshot of the applications present on a device. Of course, this needs to be done repeatedly throughout the life of the device to ensure that when new applications are added, they fit within the corporate policies and guidelines. Application inventory can also help you determine whether a device has some of the required applications — important information that you can use to determine which applications to provision to a particular device.

Blacklisting applications is a critical step toward ensuring that unwanted applications don't find their way onto employee devices. An increasing number of MDM solutions allow you to blacklist unwanted applications and prevent users from installing these applications on their mobile devices.

You need to be careful about how restrictive your application download policies are. Because many of the devices making their way onto corporate networks are consumer- or employee-owned devices, it can be difficult or impossible to enforce such policies. Even for company-owned devices, it can be a challenge to restrict usage on these devices to only corporate-approved applications.

Finally, the ability to remove unwanted applications that have found their way onto devices might be something that you require from your MDM solution. This is exactly as it sounds: the ability to remove applications from a device based on the application inventory that the device is reporting to the MDM solution.

Case Study: AcmeGizmo Application Control Deployment

Returning to our ongoing case study on AcmeGizmo, Ivan, the IT manager, is now determining the various policies he will implement for mobile device management. His view is that with the appropriate choices here, he will be able to simultaneously strengthen the security of the AcmeGizmo mobile device deployment and increase user productivity.

Your password, please

Ivan's first task is to implement a password policy for AcmeGizmo's mobile device deployment. In the security policy that he created (as discussed in Chapter 4), he specified this: "All devices must have a lock timeout of 10 minutes or less, as well as a lock password with a minimum of 6 characters. All passwords must contain at least 1 non-alphanumeric character." The enforcement policy that Ivan ended up creating is a bit more complex than that, as shown in Table 5-1.

Table 5-1	AcmeGizmo Password Policy
Policy	*Required Values*
Require password	Yes
Require alphanumeric	Yes
Minimum password length	6 characters
Minimum number of non-alphanumeric characters	1
Maximum password age	1 year
Auto-lock device	5 minutes
Password history	5
Maximum number of failed attempts	10

The password policy that Ivan implemented has the following elements:

- ✔ **Required password:** All mobile devices on the AcmeGizmo network are required to have an unlock/power on password.

- ✔ **Required alphanumeric character:** Passwords must have at least one letter.

- ✔ **Minimum password length:** All passwords must be at least six characters long.

- ✔ **Required nonalphanumeric character:** All passwords must contain at least one symbol.

- ✔ **Maximum password age:** Every password must be changed at least once per year.

- ✔ **Autolock device:** All devices must be set to automatically lock after five minutes of inactivity.

- ✔ **Password history:** A user must not reuse any of his or her last five passwords.

- ✔ **Maximum number of failed attempts:** After 10 incorrect attempts at a password, the disk on the device will be wiped (deleted) automatically. A device wipe typically does a "factory reset" of the device itself, removing all data except what was on the device when it was purchased.

Ivan has implemented all of these policies through AcmeGizmo's mobile device management solution, across the entire breadth of devices connecting to AcmeGizmo's network.

Network settings

Ivan recognizes that being able to connect to the appropriate networks not only improves productivity, but also increases the security of the AcmeGizmo deployment. As a result, he uses the AcmeGizmo mobile device management solution to preprovision the corporate wireless LAN SSID (service set identifier) and WPA2 settings on every device. Through this mechanism, when users enter the corporate offices, their devices will automatically connect to the AcmeGizmo Wi-Fi network, AG_Wireless.

Ivan has not leveraged this solution to provision VPN profiles. In Chapter 7, we describe AcmeGizmo's SSL VPN deployment using Junos Pulse. With this solution, users have an application deployed on their smartphone devices that connects them to the VPN as needed, so Ivan doesn't need to remotely set the VPN in the native device settings.

Other settings

Ivan has taken care to minimize the number of restrictions he is placing on each device. He knows that these devices are owned by the end users, not by AcmeGizmo, so he's concerned about user backlash, should his policies be too restrictive. That said, he does have security concerns to keep in mind, so he has implemented the following additional configurations on each device:

- **Restricting removable media access:** His corporate security policy states that users should not permanently save sensitive corporate data to mobile devices, primarily because there have been several issues in the past, including when Ed from Engineering recently left the company's next-generation widget designs on a device that ended up being stolen. So Ivan has taken the step of restricting access to removable media because media such as SD cards make it much easier for users to lose or steal corporate data.

- **Blacklisting applications:** Ivan implemented a mobile device management solution that allows AcmeGizmo to remove applications that may be cause for security concerns. He has already blacklisted a number of known phishing applications that he has seen in the news, and he plans to continue monitoring the application inventory across the AcmeGizmo deployment to verify, to the extent possible, potentially malicious applications. As we discuss in Chapter 10, Ivan plans to leverage anti-malware functionality in his Junos Pulse deployment to remove the overhead associated with tracking these potentially malicious applications.

✔ **Mail server access:** Several users had previously called Ivan to try to figure out how to connect their mobile devices to the corporate e-mail server so that they could download their mail, calendar, and contacts. Despite having created a very clear instructions document, Ivan was still getting these calls. As a result, he decided to preprovision access to the corporate Microsoft Exchange e-mail server on each device.

Application provisioning

As of right now, Ivan leverages his mobile device management solution to provision a handful of applications to mobile devices.

Ivan has started to take a look at a number of vendor products that offer a corporate application store. Moving forward, this is something that is of particular interest to Ivan. He has not yet chosen to implement such a solution, but recognizes that several business units within AcmeGizmo are requesting that he provision more and more applications to corporate devices. His view is that an employee self-service portal will make it easy for employees to figure out exactly which applications they need in order to do their jobs productively.

Chapter 6

Conforming to Corporate Compliance Policies

*M*any employees bring their personal smartphones to work and ask IT for help setting up VPN access. Often, corporate executives purchase the hottest devices on the market and want to use them at work. This is a change from a few years ago, when only corporate-approved devices — usually BlackBerry devices — were used to send and receive corporate e-mails.

This influx of new devices brings interesting challenges to both the device users and IT administrators. Enabling secure access on the latest devices necessitates creating policies that allow users flexibility without compromising the security of your network, and then enforcing those policies. The policies should be flexible enough to keep pace with the technology and plethora of new devices hitting the market.

In today's world of increased mobility, the issue of enforcing corporate compliance on mobile devices is more critical than ever before. This chapter discusses the process of designing corporate compliance policies. We show you what it takes to allow smartphones and other devices to be granted access to corporate data and applications, while ensuring that they adhere to certain corporate compliance policies. We also give you an example corporate compliance policy comprising various rules and policies to be enforced on smartphones.

Which Devices Are Personal, and Which Are Corporate-Owned

Most companies have a combination of personal and corporate-owned mobile devices being used within their work environments. Many employees still carry their corporate-assigned BlackBerry devices to work, for checking e-mail. Others employees have abandoned their corporate devices and purchased their own iPhone, iPad, or Android device (or all of the above).

The foremost decision for corporate compliance should be whether a device is personal or corporate-owned. For personal devices, one set of policies should be enforced, and for corporate devices, another set of policies should be enforced. Enterprises look to deploy a set of enterprise apps onto corporate-owned devices and *lock down* the configuration of the device to the extent the device operating system allows. Locking down corporate devices may include preventing access to certain utilities like the app store. With personal devices, enterprises typically cannot lock them down to the same extent. The priority for personal devices is to ensure that the devices connecting to the corporate network are protected from malware, viruses, and other threats. Enterprises also seek to deploy a sandboxed environment for their corporate applications so they're shielded from the rest of the personal data on the device.

For corporate-owned BlackBerry devices, most IT departments have infrastructure or applications installed that manage critical policy enforcements. For example, the BlackBerry Enterprise Server manages policies for BlackBerry devices, which is the reason why those devices are often the corporate standard. For employees' personal devices, however, it's difficult for IT to enforce the same set of policies on the devices, which may differ in device type, operating system, version, and vendor type. Therefore, we recommend classifying personal devices into one category and enforcing one set of policies for them, which may or may not match the policies enforced on corporate devices.

Personal devices are purchased by the employees themselves, and therefore, employees expect a greater degree of flexibility and less employer control than on corporate devices. You can expect personal devices to have the employee's personal content, including photos, videos, SMS messages, and other private data.

Now let's look at what policies you want to enforce on each category of devices. Here's how your classification of devices may look:

✔ **Personal devices allowed for corporate access:**

- Android devices running Android version 2.1 or later (like the HTC Desire and Samsung Nexus S)
- Nokia E7 and other phones running Symbian 3
- Apple iPhone 3GS, iPhone 4, and iPad running iOS 4.0 or later

✔ **Corporate-assigned devices allowed for corporate access:**

- BlackBerry (all models)

✔ **Devices that are *not* allowed into the corporate network:**

- Android devices running Android version 2.0 or earlier
- Windows Phone 7 devices

Of course, your policy may look different from this example. Choosing devices to allow into your corporate network depends largely on your tolerance for risk and device heterogeneity. So make sure to research the latest smartphone platforms available today and determine which ones you might allow into your network.

 From a compliance perspective, we recommend rejecting jailbroken or rooted devices from entering your network, and stating that in your policies. The risks posed by such devices to your corporate network far outweigh the flexibility of users being able to install private apps, outside of the application stores. Also, such devices are unsupported by the device vendors, so any issues with the phones will not be resolved by the vendors. For more information on jailbroken or rooted devices, see Chapter 2.

Setting Passcodes on Mobile Devices

Setting passcodes on mobile devices is the most basic security requirement for any smartphone to be allowed into a work environment. Passcodes require the user to enter a passphrase to unlock the phone. Devices can also be configured to lock automatically after a configurable timeout period. (Typically, five minutes is ideal.)

From a compliance perspective, take a look at the passcode policies that you may want to enforce on smartphones:

✔ The device needs a passcode configured.

✔ The passcode needs to be of a certain strength, incorporating at least one digit or complex character.

✔ The passcode needs to expire after a certain time period.

✔ The device should lock after a certain time period of inactivity.

✔ Some sort of action should be taken if the threshold for failed attempts to enter the right password (such as ten consecutive bad passcodes entered) is reached.

For different organizations, the exact passcode requirements will vary. For many, it might suffice to simply require a passcode on each smartphone in the corporate network. For others, it might be necessary to enforce additional restrictions, such as the passcode strength and expiry time period. What you specify for your organization's passcode requirements largely depends on your tolerance for risk and adherence to other corporate policies or restrictions.

At this time, you also need to decide whether to enforce the same set of passcode policies on both personal devices and corporate-owned devices. As we explain earlier in the chapter, you have the liberty to define different compliance policies for corporate-owned and personal devices and establish different passcode policies for the two categories of devices.

Tables 6-1 and 6-2 summarize what a portion of the compliance policy might look like.

Table 6-1   Compliance Policies for Allowed Device Types

Personal Devices	*Corporate-Owned Devices*
Android devices running version 2.1 or later	BlackBerry (all models)
Symbian 3 devices	
iPhone 3GS, iPhone 4, and iPad running iOS 4.0 or later	
No jailbroken or rooted devices	

Table 6-2	Compliance Policies for Passcodes
Personal Devices	*Corporate-Owned Devices*
Need a passcode	Need a passcode
Passcode strength (for example, it should be at least 8 characters long and must include at least one digit)	Passcode strength (for example, it should be at least 8 characters long, and must include at least one digit)
Passcode expiry	Passcode expiry
Time before autolock	Time before autolock
Action taken upon 10 unsuccessful attempts	Action taken upon 10 unsuccessful attempts

Encrypting the Contents of the Device

Data encryption prevents sensitive smartphone data from being accessed without entering the device owner's passphrase or secret key. *Encryption* refers to the process by which vital data is made inaccessible to users who don't know a secret phrase or password. For example, the passphrase used to lock the phone could be used as a password to encrypt the data on the device.

Typically, smartphones are used within the workplace to access corporate e-mail, browse intranet web pages, or even access client-server applications like Oracle or SAP. In addition, the devices may also store contacts, SMS messages, or files related to work.

From a compliance perspective, encryption of the device should require such content to be encrypted before being allowed into the corporate network. Here's a list of the types of data that should be encrypted on the device, at a minimum:

- E-mail
- SMS messages
- Contacts
- Calendar
- SD card contents, which may include files of various types

The challenge in requiring encryption to be enabled on smartphones is that not all smartphones support hardware encryption. For example, the Apple iPhones and iPads support it on devices running version 4.0 or later, but not on earlier versions. Android devices don't have encryption capabilities, at the time of writing this book.

The specific policies for data encryption will vary for different organizations, depending on each organization's tolerance for risk and the importance of this particular form of security.

At this time, you should also decide whether to enforce your encryption policy on both personal and corporate-owned devices. Ideally, you'd want to do so, thereby ensuring that the policies are consistent for all devices in the corporate network. However, because some devices (we're looking at you, Android) don't currently support encryption, you will need to decide whether to let some devices into your network without encryption, or enforce encryption as a policy throughout the smartphone network, thereby denying access to some popular Android devices today.

Table 6-3 summarizes the encryption policy requirements for smartphone compliance.

Table 6-3	Compliance Policies for Device Encryption
Personal Devices	**Corporate-Owned Devices**
May be identical to the policies for corporate-owned devices.	Encryption must be enabled for all the following data types: e-mail, SMS messages, contacts, calendar, and SD card contents, including videos and photos.

Requiring VPN on the Device

Virtual private network (VPN) refers to the secure connectivity between a device and a VPN gateway or server installed within the corporate network. When a VPN tunnel is established between a device and the VPN gateway, all communication over that tunnel is encrypted. This encryption provides security for data being exchanged between the device and the corporate network.

Hackers can snoop on data that isn't encrypted as it's on its way to the device. For example, it's possible for a hacker at a café to snoop on unencrypted data being received on another person's smartphone. This is why you want end users to connect via VPN when they're accessing corporate data in public places.

You may have used a VPN to connect to your corporate network from your PC at home. Similar technology is available for several smartphone devices. A *VPN connection* encrypts the data communication from and to the smartphone, thereby making it impossible for hackers to intercept and steal the data being exchanged.

So, the most critical requirement of data encryption is to enforce VPN access as a compliance requirement. If you are an IT administrator, that means enforcing VPN for all smartphone users to connect to their work e-mail or other applications. Most VPN vendors like Cisco and Juniper have VPN solutions available for some or all smartphone types.

Enforcing VPN on smartphone devices requires you to have a VPN server or gateway installed in your network. The devices need to connect to the server when setting up the VPN tunnel.

Here are the VPN policies you may want to enforce on smartphones:

✔ Allow users to check corporate e-mail, browse intranet pages, and/or use client-server applications.

✔ Enforce strong authentication on the devices, including one or more of the following types:

   • Username and password

   • Certificate-based authentication

   • One-time password (passwords expire after just a single use)

✔ Manage a single set of policies to set consistent VPN policies for not just smartphones, but also Windows and Mac computers.

Certificate-based authentication and one-time password authentication require you to deploy certificates to smartphone devices as well as set up infrastructure to configure the one-time password server in-house. Be sure to look up the vendor documentation for deployment guides and instructions.

Finally, VPN — or, in general, secure connectivity from smartphones to the corporate network — may differ from corporate devices to personal devices. For example, BlackBerry devices maintain a secure connection to the BlackBerry Enterprise Server that is typically installed within a corporate network, which saves you from needing a VPN. For all other smartphone types, you're better off requiring and enforcing a strong VPN policy.

Table 6-4 summarizes the VPN requirements for smartphones from a compliance perspective.

Table 6-4	Compliance Policies for VPN
Personal Devices	*Corporate-Owned Devices*
Smartphones should support VPN from the device to a VPN gateway installed within the corporate network. Enforce your favorite vendor's VPN support on each personal smartphone allowed into your network.	For BlackBerry devices, we recommend deploying the BlackBerry Enterprise Server, which sets up a VPN-like secure connection to the smartphones.
Enforce granular application access. Allow any or all of the following applications: e-mail, intranet, and client-server applications.	For other types of devices such as the Apple iPhones, iPads, or Android or Windows Mobile devices, we recommend deploying an IPSec or SSL VPN to manage corporate access.
Enforce strong authentication methods.	

Protecting the Device from Viruses

Because we're talking all about compliance in this chapter, take a look at the various aspects of smartphones that are vulnerable to hackers, and what you can do to protect the devices and data on them:

- ✔ **Malicious apps:** Certain apps can steal information from the device and relay it back to a hacker's server. Information that can be stolen includes the contacts, calendar, messages, and other content stored on the device. Several apps prompt users to allow them to access their GPS location, for example. Allowing GPS access to an app provides crucial information about where the device and its owner are at any point in time, putting data at risk because the device could be physically stolen. It's critical to monitor the behavior of apps and weed out the ones that are malicious.

- ✔ **Spam:** Mobile devices are susceptible to receiving spam in several forms, including text messages, instant messages, and e-mail, and via online games. These are all ways in which spammers target smartphone owners. The spam messages are typically solicitations for products or services, often fraudulent.

- ✔ **Worms, viruses, and Trojans:** Just like on Windows PCs, software viruses can affect smartphones and replicate by sending copies of themselves to all contacts found in the address book. Devices can receive such files via SMS, MMS, e-mail, Bluetooth, or any of the plethora of communication methods available for smartphones.

From a compliance perspective, here's a list of items that you should consider enforcing on smartphones in your corporate network:

- ✔ Comprehensive antivirus protection, with automatically updated virus signatures to protect against Trojans, worms, and other threats

- ✔ Antispam protection, with the ability to automatically delete spam

- ✔ Antimalware protection, to detect malicious apps that suspiciously track user information via GPS

- ✔ Firewall protection, to set traffic filters that control the traffic flowing into as well as out of the device.

Not having such software is akin to letting your users connect to your corporate network from computers that have no security software (like antivirus or antispyware). You'd never let that happen, so why allow smartphones to be able to connect without similar protection?

As for personal versus corporate-assigned smartphones, you should enforce the virus protection on both types of devices, just like you would on both home computers and corporate laptops.

Applications like antivirus protection usually affect the battery life of a smartphone. Be sure to analyze the effects on battery life when you shop around for smartphone antivirus solutions.

Most antivirus software products work off virus signatures that are regularly updated by the vendor. Look for solutions where the virus signatures are automatically updated from the vendor to each smartphone. You can't expect the smartphone user to manually update the virus signatures.

Table 6-5 summarizes the compliance policies to protect smartphones from viruses and other threats.

Table 6-5   Compliance Policies for Protection against Malware

Personal Devices	*Corporate-Owned Devices*
Comprehensive protection from viruses, Trojans, worms, malware apps, and spam	Identical to the policies for personal devices
Firewall policy enforcement	

Protecting the Device from Loss and Theft

A critical policy for smartphone compliance is the ability to take actions when a smartphone used for corporate access is reported lost or stolen. Employees carry critical information on their smartphones, including personal and corporate e-mail, contacts of people at work, SMS messages, and so on. When an employee loses a smartphone, such information is liable to being stolen. Therefore, it's extremely important to take immediate action when a device is reported lost or stolen.

Here are the kinds of actions that you can take to mitigate the risks of the loss or theft of a smartphone:

- ✔ Locate the device via the GPS location.

- ✔ Remotely lock the device so that others can't access data on it unless they know the password.

    One way to mitigate the threat of somebody guessing the user's passcode is to set a limit for the number of incorrect login attempts so that after maybe five or ten attempts, the device is automatically locked. Alternatively, you could temporarily suspend the user's authentication until the user calls the help desk to unlock the account.

- ✔ Remotely set off an alarm so that the theft of the device becomes obvious to others in its vicinity.

- ✔ Remotely wipe the contents of the device so that no traces of personal or corporate data remain on it.

- ✔ Remotely lock or wipe the device as soon as the SIM card on the device changes. (If the SIM card changes, it's an indication that the thief is attempting to reuse the phone.)

Each of these actions mitigates the risks of losing sensitive data on lost or stolen smartphones. You should also evaluate whether you need these actions to be taken by you or the employees themselves.

In the case that employees can take such actions themselves, they would need to log in to a web portal to authenticate themselves with a username and password. Once authenticated, they would take any or all of the actions discussed here on their phone. This kind of model allows employees to take immediate action on their lost or stolen phones.

On the other hand, if you (corporate IT) choose to get involved, the employee would need to call the help desk to report a lost or stolen phone. The help desk would retrieve details of the phone from the phone number provided by the employee and then take any of the actions discussed here.

It's important that such actions are taken as soon as possible after the device is reported missing. Delaying actions such as remote wipe or remote lock increases the risk of sensitive data getting stolen from the missing device.

The definition of *remote wipe* has subtle differences for different mobile platforms and vendors. For example, on some platforms, a remote wipe indicates that all user content is removed from the device, leaving it in what is called a "factory-default" configuration. Some vendors can wipe selective content from the device, removing enough data to prevent confidential data from getting into the wrong hands.

From a compliance perspective, this policy should be enforced as much on personal smartphones as on corporate-owned ones.

Table 6-6 summarizes the loss and theft compliance policies to be enforced on smartphones for corporate usage.

Table 6-6   Compliance Policies for Loss and Theft Protection

Personal Devices	Corporate-Owned Devices
Locate via GPS	Identical to the policies for personal devices
Remote lock	
Remote alarm	
Remote wipe	

Managing Devices at Scale

In a small- to mid-sized company, as many as several hundred smartphones might connect to the corporate network every day. With larger companies, the number of smartphones in the corporate network can easily be in the thousands or tens of thousands. In such cases, it's critical to manage the policies that are deployed on these devices for compliance purposes.

Here are some considerations for evaluating the compliance needs for the management of smartphones in a corporate environment:

- **Management at scale:** Whatever management process or system you use must scale for thousands of devices. Remember to estimate for more than you need today because, as your organization grows, the number of smartphones in your network will grow just as fast, if not faster. The management system must be able to deploy all compliance policies to smartphones in a centralized manner. Seek a single solution, or a single vendor, to offer a centralized solution that can manage all types of mobile devices and smartphones from a single console.

- **Centralized inventory management:** The centralized management console must be able to report an inventory of mobile devices managed within the corporate network. This capability should also include the ability to report the inventory by device type, vendor, and operating system. This allows you to pull up reports for your bosses to show the number and types of smartphones connecting to the corporate network. In a rapidly evolving smartphone market, it's important to keep an eye on the trends of devices in your network.

- **Centralized logging and reporting management:** The centralized management console must also be able to generate logs and reports of incidents, as well as compliance and policy violations in the network. For example, you must be able to run reports that show the number of virus infections that were immediately caught in the last 30 days, or the number of new devices that connected to the network in the last week or month. This must also provide details of actions taken on a particular smartphone.

- **Notifications:** The centralized management console must provide real-time notifications to IT staff when a critical event happens. For example, virus infections must be reported via SMS, e-mail, IM, or other means to interested subscribers. Though the virus infections might be immediately caught and corrected by the software, it's important to have the framework available to report such events in real time.

- **Configuration management:** Another part of managing smartphones is the management of configurations and versions of each device. This may be optional to some companies, who state that only certain OS versions are supported within the network and that device owners are responsible for upgrading the device to a supported version.

Most of the functions we describe in this section are provided by the BlackBerry Enterprise Server for BlackBerry devices only. You essentially have to look for a single solution that does a similar set of functions for all the other devices out there, including those from Apple, Google, Samsung, Motorola, and others. You may be better off retaining the BlackBerry Enterprise Server to enforce compliance on BlackBerry devices and using a separate solution for all other device types.

Another aspect of managing configuration might be to push settings, policies, or applications as a configuration update to the devices. If your company has proprietary apps to install on your employees' smartphones, you need a centralized configuration management system that can manage the deployment of policies and software to smartphones.

It's important in this case, too, to ensure that the configuration management console is centralized in a single console for all device types. It's ideal to have a single centralized console, or at least minimize the number of systems that can manage your smartphone diversity in the corporate environment.

Table 6-7 summarizes the smartphone management requirements for corporate compliance.

Table 6-7	Compliance Policies for Centralized Management
Personal Devices	**Corporate-Owned Devices**
Employ some or most of the policies that apply to corporate-owned devices.*	Management at scale for thousands of devices
	Centralized inventory management
	Centralized logging and reporting
	Real-time notifications
	Centralized configuration management

* Note that it may not be possible to force personal devices to upgrade to a certain OS version, or push out corporate software or settings to them. In those cases, select the appropriate policies to enforce for personal devices.

Backing Up the Contents of the Device

Regular backups of smartphone contents are just as important as backing up the contents of Windows or Mac computers. From the user's perspective, regular backups are extremely useful when a device is lost or stolen because the user can easily restore his lost device's contents to a new device. From a corporate perspective, regular backups provide checkpoints that provide insight into the data and contents of smartphones for potential forensic analysis.

Several smartphones available today have a large amount of space on their hard disks, often in tens of gigabytes. For example, 16GB and 32GB configurations are common for most smartphones. If it's difficult to back up that volume of data for thousands of smartphones, it might be useful to back up only the critical data that resides on them.

From a compliance perspective, here are the types of data that should be backed up from smartphones:

- Contacts
- Calendar
- Call log
- SMS messages
- Photos and files (if needed)

This particular policy may also differ for corporate-owned devices, in comparison to personal devices. For example, you may not want to enforce backups of personal files, including videos, pictures, and messages from personal smartphones. On the other hand, it may be acceptable to back up more data from corporate-owned devices. Depending on your organization's particular tolerance for risk, you can choose to back up any or all of the data we just listed.

When you devise a smartphone backup policy, be sure to think through the following aspects as well:

- **Where is all that data stored?** You can choose from several cloud-based solutions, which back up all the smartphone data to a central system in the vendor's cloud. If this isn't acceptable for your company either due to geographical or industry restrictions, insist on a solution that stores all the backed-up data on a server within your corporate network.

- **Who is authorized to see that data?** A typical backup solution for smartphones must involve the device users to invoke a backup operation manually whenever needed, or schedule backups at periodic intervals determined by the IT administrator. An IT administrator could schedule a backup every day or every week, or set any other schedule that seems reasonable.

    Typically, the user can restore the contents to the device (or a replacement) manually without needing a third party (an IT administrator) to intervene. The user's backups should be protected by a user-configured passphrase.

    In addition, you may also want to give administrators access privileges to that backed-up data. In case users forget their passphrases, it should be possible to reset the passphrase, just like resetting Active Directory passwords.

Depending on where the user's information is stored, it is important for the software vendor to assign permission to appropriate parties to see that sensitive information. For example, even the mobile device management vendor's customer support group may not be granted access to the user's photos, contacts, or SMS messages. It is important to check whether the vendor's software can assign granular privileges to various groups of users to see sensitive information that belongs to mobile end users.

It's important to identify the list of people who are authorized access to all the backed-up data stored either within your corporate network or in the cloud of your vendor's network.

✔ **Are communications encrypted during backups and restorations**? You should explore your vendor's solution to check how the smartphone's contents are backed up. The data must be encrypted back and forth from the device to the central backup server. If that communication isn't encrypted, it's possible for hackers to snoop in on that traffic and access the data being backed up (or restored).

Table 6-8 summarizes the backup compliance policies for smartphones.

Table 6-8	Compliance Policies for Regular Backups	
Compliance Policy	*Personal Devices*	*Corporate-Owned Devices*
Regular backups of smartphone contents	Some, maybe not all, of the data that is backed up from corporate-owned devices. At a minimum, include the following: contacts, calendar, and call log.	Contacts Calendar Call log SMS messages Photos and files (optional)
Regular backup policies	All policies applicable to corporate-owned devices	Location of backed-up data Access control to backed-up data Encryption of backed-up data

Monitoring and Controlling Contents of the Device

Depending on your corporate policies regarding security and risks, you may be required to inspect and tightly control data and applications residing on corporate-owned devices. This includes Windows and Mac computers as well as smartphones and other mobile devices. For example, it's fairly common in the government to have tight policies controlling what users can access from their computers and mobile devices.

In this section, we explore what policies might make sense if you're required to comply with strict requirements controlling the data and applications on your corporate-owned mobile devices.

Here are the various types of contents on smartphones that are liable to being controlled for compliance purposes:

- ✔ SMS messages
- ✔ Contacts
- ✔ Photos
- ✔ Videos
- ✔ List of apps installed

Controlling these types of data involves sending copies of the data to the central management console. From that console, an administrator can analyze the types of messages being received on devices that are used within the corporate network, or the photos that are being taken on them, or the applications that are installed on them.

You can restrict the apps installed on smartphones within the work environment to a list of corporate-approved apps. For example, your company may decide that the Facebook app should not be installed on any smartphone that is used for work-related purposes, such as checking e-mail.

For compliance purposes, any of the previous types of data could provide forensic information, if it were needed to analyze the causes of a data breach. But again, it might not be required (or even legal!) in your company, depending upon the industry type, geographic location, and federal or local laws.

In addition, the nuances of personal and corporate-owned devices may provide a distinction between the types of content that are allowed to be inspected. For example, in many companies in the U.S., employees would balk at their employer inspecting and controlling their photos, videos, and contacts. This is an example of a compliance policy that may differ in nature, depending on the type of employer, industry, or location of the company.

Table 6-9 summarizes the monitoring and control policies for smartphone compliance in a corporate environment.

Table 6-9	Compliance Policies for Monitoring and Control
Personal Devices	**Corporate-Owned Devices**
Some, maybe none, of the policies enforced on corporate-owned devices	Some or all of the following content may be inspected, depending upon your company's need: Contacts SMS messages Photos, videos, and apps installed

Case Study: AcmeGizmo Compliance Requirements

On a typical day, any number of employees will have purchased new, personally-owned devices and want to get access to AcmeGizmo corporate data and applications from those devices. Consider a scenario where Fred from the Finance department has recently purchased a new Samsung Galaxy S II Android smartphone and wants to access his e-mail and a few select applications from that device. Follow along as Ivan the IT manager determines whether he can approve this device and user for network access.

Operating system compliance

Ivan's first step is to determine whether this new device is in compliance with the company's list of approved devices and operating systems. The policy states that all Android devices must run version 2.2 or newer. Because Fred's device is running Android OS version 2.3, Ivan is satisfied that it meets AcmeGizmo's policy.

Password compliance

Ivan also needs to ensure that this device has an appropriate password policy. Rather than checking the device configuration, Ivan deploys a policy to the device via AcmeGizmo's mobile device management solution.

Fred receives an e-mail with instructions to browse to a site and perform some registration-related tasks, and then the device management profile is deployed to his new device. This profile configures the password policy restrictions on Fred's device, ensuring that he sets a lock password that meets with AcmeGizmo's corporate security policy.

Encryption compliance

Ivan must also check to ensure that the device has encryption enabled. This particular device, the Samsung Galaxy S II, has been targeted toward the prosumer (professional consumer) and corporate markets, and it has native encryption functionality (even though the base Android 2.3 operating system doesn't).

Ivan simply uses the same mobile device management profile deployed to adjust the password settings to ensure that the encryption on this device is enabled. In fact, the encryption and password settings are applied to all devices accessing the AcmeGizmo network, so no additional work is required for Ivan to accomplish this.

VPN and endpoint security compliance

A few additional tasks are required before the device is in compliance. Because Ivan has selected Junos Pulse as AcmeGizmo's device management and security solution, Fred downloads the Junos Pulse application from the Android Market. This solution protects the device through antivirus and other device security features while simultaneously providing the VPN component and configuration, which allows Fred's smartphone to securely access the AcmeGizmo network.

This client also leverages the multifactor authentication solution that AcmeGizmo has put into place, ensuring that Fred authenticates to the network properly, using a one-time password solution rather than a simple username and password. This increases security by ensuring that a stolen password and device don't allow a malicious entity to gain access to sensitive corporate data.

Loss and theft protection

Ivan's last step is to ensure that Fred's device is appropriately protected from loss and theft. Because Ivan has already deployed the Junos Pulse device security solution onto Fred's device, this protection exists. If Fred loses his new smartphone or if it is stolen at any time, he simply reports the loss to AcmeGizmo IT, and all sensitive data can quickly and easily be removed from the device.

Part III
Securing Smart Device Access

The 5th Wave   By Rich Tennant

"Hold on Barbara. I'm pretty sure there's an app for this."

In this part . . .

Other parts of the book help you develop and implement policy, but this part moves on to the real world — which eventually means your network. How do you build the system of seeing, accepting or rejecting, or limiting access to the hordes of devices entering your main branch and remote offices each and every hour?

Chapters 7 and 8 contain all sorts of old friends to help you in your tasks: friends like IPSec and SSL VPNs, Wi-Fi, user authentication, and application access control.

And just to make sure you and your old friends are ready to party, our book's model company, AcmeGizmo, has a case study to put you in the mood.

Chapter 7

Securing Data in Transit with VPNs

*I*f your organization is like countless others that we've worked with, you've spent a great deal of time and money over the past several years developing policies and implementing solutions to allow for secure remote access. The ability to work remotely while retaining access to corporate applications has changed when and where people work, resulting in untold productivity gains for everyone from traveling employees (such as executives and salespeople) to folks looking to squeeze a bit more work from their day in the evenings and weekends.

As time has progressed, you've no doubt built the infrastructure that allows these levels of productivity, without sacrificing security. The cornerstone of most remote access policies involves use of a virtual private network (VPN) — typically either an SSL (Secure Sockets Layer) VPN or an IPsec VPN — though there are usually other technologies, such as endpoint security solutions and strong authentication technologies, involved as well.

Your VPN purchase decision was probably driven by the need for employees to access the work network from their corporate-owned and -managed Microsoft Windows computers. At most, you might allow them to also access some amount of corporate data via their own PCs or kiosk machines when traveling. Flash forward to today, where you are no doubt hearing overwhelming demand to access these same systems from a range of new mobile devices. Unfortunately, most remote access solutions on the market today do not support mobile devices. If you're lucky enough to have chosen a VPN solution that fully supports a range of mobile device platforms, you have a step up on some of your peers in the IT world, though possibly there's still much work to be done to establish best practices and policies for mobile device access inside your organization. If your VPN solution doesn't support mobile device platforms, fear not; plenty of vendors are building products for exactly this purpose, and their salespeople are no doubt already beating down your door asking you to take out your checkbook.

This chapter focuses on remote access technologies that you can use to secure mobile device access to the corporate network. It explores user identity and machine compliance and their role in developing a secure remote access strategy for mobile devices in your network.

Comparing IPSec VPNs and SSL VPNs

Two types of VPNs represent the majority of global remote access use cases: IPsec and SSL. Depending on what your vendor provides, and your company's policy requirements, either type might work as you extend remote access to smartphones.

To understand the similarities and differences between IPsec and SSL VPNs, you need to understand VPNs in general. VPNs allow ways to transmit sensitive data across shared networks without it being intercepted or stolen. VPNs were initially designed to service site-to-site networks. Before the availability of VPN solutions, organizations relied on expensive, leased point-to-point data circuits, such as T-1 lines leased from the major telecommunications providers, or shared but still relatively expensive technologies such as Frame Relay. By using a VPN, organizations can enjoy the benefits of shared networks, without the security concerns that are typically associated with transmitting data over the Internet. A VPN provides encryption for traffic as it traverses the Internet, ensuring that this traffic is just as secure as if that traffic were to traverse a separate point-to-point connection.

Site-to-site VPNs are responsible for authentication (identifying users or machines attempting to establish a VPN connection), encryption (to ensure that any intercepted traffic can't be read), and integrity mechanisms (to ensure that traffic isn't tampered with while in transit).

Over time, VPNs were adapted to be used by remote workers. When applying VPN to remote-worker use, many of the concepts and protocols from site-to-site VPN connections remain the same: authentication, encryption, and integrity mechanisms.

IPsec and SSL VPNs make up the majority of today's enterprise remote access deployments. Here is how the security protocols work for each type of VPN:

✔ **IPsec VPNs** provide a secure, network-layer (Layer 3) connection to the corporate network. As data traverses the Internet from the mobile device to the VPN gateway, it is encapsulated and encrypted. After the traffic passes through the VPN gateway and onto the LAN, it is no different from traffic coming directly from end users on the LAN. The result is access that is very similar to access that a user would get when physically connected in their own office: full connectivity to all resources and applications. Of course, this level of access isn't without its disadvantages.

For example, your organization might not want to allow a user on a mobile device access to all enterprise applications if all that user really needs is the ability to check her e-mail from her mobile device. By limiting access to specific applications, you can control the potential risks associated with providing more complete access from a compromised or insecure machine.

✔ **SSL VPNs,** the type of VPN most commonly deployed for new enterprise remote access deployments, can almost always provide the same Layer 3 VPN capabilities that are provided with IPsec VPNs, while also providing the additional control necessary to restrict access for users or groups of users. As an example, a user attempting to access the corporate network from a company-owned Microsoft Windows laptop, with all the required security fixes and patches, might be an ideal candidate for full Layer 3 access to the corporate network. That same user attempting to access from his personally owned Apple iPhone, on the other hand, might be subject to stricter controls that allow him access to only a few selected web-based applications and e-mail. An SSL VPN allows for this additional control and enables you to prohibit more permissive access, should you deem it necessary.

Of course, whether you have an IPsec VPN or an SSL VPN, platform support is a key requirement. Not every vendor supports every mobile platform available, so it's a good idea to work with the vendor of your VPN gateway to determine whether the existing product supports the types of mobile devices that you plan to provide access to. In many cases, the vendor needs to build and support a client application on the endpoint device; this challenge really adds up across several popular platforms, so the vendor might not support everything out there.

Validating User Identity for VPN Access

Before you allow access to the corporate network from any device, you should first identify the user attempting to access the corporate network. Organizations typically view user identity validation as two distinct pieces: user authentication and user authorization.

Here is a description of user authentication and user authorization:

- ✔ **User authentication** involves validating that a user truly is who she says she is. In other words, user authentication proves that the person attempting to log in to the VPN as SueB really is Sue Berks, and not Joe Hacker.

- ✔ **User authorization** is another important part of the user identity process. User authorization typically involves determining a user's role or job function in the organization, for the purpose of granting him access to a particular set of data and applications.

In the sections that follow, we take a closer look at both of these areas of user identity validation.

Authenticating VPN users

From an authentication perspective, most leading VPN offerings provide integration with a mix of standards-based and proprietary authentication servers, such as the options discussed in the following sections.

As with many security technologies, a range of security strengths are offered through these various solutions. Organizations that are very security conscious typically use a strong authentication solution such as a one-time password system or X.509 digital certificates. The use of strong authentication has become very popular in recent years; we recommend it as a best practice for all organizations. Less security-conscious organizations stick with static username and password systems for remote user authentication.

If you deploy systems using static credentials, we strongly recommend the use of password policies and user training that minimizes the risk of stolen or easily guessed passwords. Password policies include enforcing a minimum password length and a certain number of numeric or special characters. We also advocate implementing technologies that prohibit the use of common words or names in passwords, force user password changes on a periodic basis, and limit reuse of prior passwords.

All the fancy technology in the world will do no good if your end users routinely place sticky notes with their passwords written on them under their computers, so constantly remind and train your end users on the risks associated with bad practices and poor passwords. Or better yet, stick with our recommendations and implement a strong authentication system.

Local authentication

Local authentication is an onboard database for authentication of users. The entire user account management and record storage is done on the VPN appliance.

Most VPN vendors offer this type of authentication, though it's used primarily for administrator authentication or for smaller organizations. Most large organizations invest in (or plan to invest in) an external user authentication solution.

Lightweight Directory Access Protocol (LDAP)

LDAP (Lightweight Directory Access Protocol) is a standard protocol for querying a directory database and updating database records. As one of the more commonly used interfaces in VPN deployments, LDAP acts as the protocol of choice for querying many types of databases, including Active Directory.

Active Directory (AD)

Active Directory is one of the leading directory servers, and most organizations deploy it, to some extent. Many VPN servers offer a native Active Directory authentication server interface, but AD deployments can also leverage LDAP/ LDAPS (LDAP over SSL) for queries and updates.

RADIUS authentication and one-time password systems

Multiple-factor authentication, such as one-time passwords (OTPs) and digital certificates (see the following section), have become very popular for remote access, replacing static usernames and passwords in many organizations. Like with digital certificates, a number of technologies have evolved that make it far easier and less expensive to activate and manage one-time password solutions, and organizations have adopted these technologies as a result of those developments.

Most VPN systems provide a standard way to interface with these OTP systems through the RADIUS protocol. Remote Authentication Dial-In User Service (RADIUS) provides authentication, authorization, and accounting services; and most OTP systems available on the market today support RADIUS. Some VPNs also provide native support for proprietary OTP systems, but you don't need this special native integration for most deployments because the RADIUS interface can provide the same functionality.

X.509 certificate authentication

In recent years, X.509 digital certificates have become more popular as an authentication method. They're issued by several trusted certificate authorities (CAs) to organizations and end users. These CAs hold the power to revoke these certificates at any time. Because they're based on secure digital certificates, this form of user or machine credential is impossible to spoof or steal without acquiring the private key, which is kept protected and never exchanged. The deployments within the U.S. Government have been a huge driver for adoption of X.509 certificates, due to mandates requiring their use not only by government agencies, but also by government contractors and other private-sector organizations. As a result, both software and hardware support have improved significantly in recent years, making deployment and ongoing administration much simpler.

When a VPN appliance supports X.509 digital certificates, that appliance must perform validation of a certificate to ensure that the certificate hasn't been revoked. The VPN validates the certificate with either

- **CRLs (certificate revocation lists):** CRLs are essentially lists of revoked certificates that are distributed by the certificate issuer.

- **OCSP (Online Certificate Status Protocol):** OCSP was introduced as a way to bypass some of the limitations of CRL checking (such as the size of the lists), and it specifies a way to verify certificate status in real time.

In addition to certificate status validation, the VPN might also retrieve user attributes from the certificate so that the VPN access control system can compare those attributes to attributes in a directory, for example, or map users to specific roles in the VPN implementation. Most VPNs also allow the administrator to specify which certificate authorities (CAs) will result in a successful authentication.

Security Assertion Markup Language

Security Assertion Markup Language (SAML) is a standard for authenticating and authorizing users across different systems. Essentially, it's a Single Sign-On (SSO) technology. Some SSL VPN appliances provide support for SAML, allowing users who are already logged in to other systems the ability to seamlessly log in to the SSL VPN system as needed. SAML authentication solutions aren't usually associated with IPsec VPNs because they are primarily a web-based authentication system leveraging Internet browsers, not an area typically associated with IPsec VPNs.

You can find a variety of identity and access-management platforms available that support SAML. In most SSL VPN deployments that use SAML, the primary use case is SSO to SSL VPN–protected resources and applications, not signing in to the SSL VPN itself. So not many enterprises have used this authentication method, in our experience.

Determining a user's role

User authentication is only one piece of the puzzle in identifying and providing appropriate access to users. Another piece is *authorization*. Authorization maps information about a user to the credentials provided at login.

In some cases, the relevant information is stored in the authentication request. For example, an X.509 digital certificate or a SAML assertion might contain information that allows the VPN to do the appropriate role mapping. In other cases, however, the VPN appliance needs to query the directory.

Many organizations have stored information, such as group membership or other attributes related to each user, in Active Directory or some other LDAP database. Upon login, the VPN queries the database to get these details and uses that information to assign the user to a specific role. For example, an LDAP query for John Doe might return information indicating that John is an employee in the Engineering group. As a result, John gains access to the intranet, corporate e-mail, and various engineering-specific resources.

Discriminating by Device Profile

Over time, many organizations have built policies that allow them to discriminate between various device types and device security posture levels in order to set an appropriate level of access for a particular session. For example, a user attempting to access the network from an appropriately protected and registered mobile device might be granted full network access, whereas a user attempting to connect from an unknown mobile device might have her data and application access severely restricted. Or that user might not be able to access the network at all until she follows the appropriate steps to make the device compliant.

In order to discriminate between devices with varying security posture levels, it is crucial to validate the endpoint machine prior to allowing the user to connect to the network in a remote-access setting.

Note that at the time of this writing, very few VPN products offer a solution to the challenges outlined in this section, but it is anticipated that additional vendors will attempt to solve these challenges for their customers in the near future. Additionally, commercially available IPsec VPN products don't provide any feature sets for device security posture profiling, so commercially available solutions are restricted only to SSL VPN products.

Figure 7-1 shows a typical scenario where access controls are applied based on the device and the device security posture. In this case, the SSL VPN policy dictates that a different level of access is granted to the end user based on whether the user's machine is in compliance with the policy, as detailed in the following list:

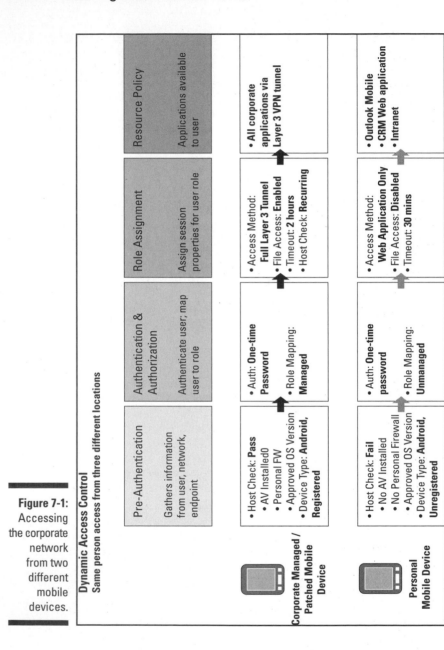

Figure 7-1:
Accessing the corporate network from two different mobile devices.

▶ **Corporate-managed or patched mobile device:** In this case, the user is attempting to access the network from an Android OS device that is registered and has been provided by the organization. Antivirus and personal firewall software is installed on the device, and the organization can remotely wipe and track the device should it become lost or stolen.

Based on this device information, and on the user's authentication with a one-time password, the managed role applies for this particular session. Because the user is coming from what appears to be a managed device that meets all the security requirements, the user is granted full, Layer 3 network access. Along with that access comes the ability to reach all applications, an experience not unlike the experience that users receive when they are in their offices accessing the network from their managed laptops.

✔ **Personal mobile device:** In this example, the user is attempting to access the network from an Android OS device, but this time, it is the user's own personal device that she brought from home. In this case, no endpoint security software is installed, and the device hasn't been registered, so the organization doesn't have the capability to remotely wipe the device or track it if it's lost or stolen.

Based on this device information, and on the user's authentication, the unmanaged role applies for this particular session. In this case, the user has access to far fewer applications and resources than she had when accessing the network from the corporate-managed device. Here, she can access only a select few web-based applications, she has no ability to access corporate file shares, and only a very short network inactivity timeout is employed to help guard against loss or theft because the IT admin can't provide these protections on this particular device.

In actuality, you will likely have several different policies for different device types, so Figure 7-1 is a more simplistic picture than what you will end up with. But it does illustrate what you can do with currently available technology.

Profiling devices and applying policies

The most commonly used and easiest to configure types of endpoint security policies are those that verify the presence, operation, and up-to-date nature of third-party endpoint security applications. These types of policies ensure that the mobile devices that you are allowing access to the corporate network have a security posture and device identity that allow you to feel comfortable allowing the device onto the network.

In many cases, your VPN vendor has already done much of the legwork for you and has created a list of predefined security policies that you can easily implement to scan for this assurance. Note that as of press time, only a few VPN vendors, all of them SSL VPN vendors, provide assurance that these types of solutions are in place. Likely, more will add such capabilities over time as smartphone adoption in the enterprise continues to increase in popularity.

Look for these common policy types provided by VPN vendors:

✔ **Device type:** Device type scans allow you to identify what type of device is connecting, or attempting to connect, to the VPN. In some cases, you simply want to restrict access to certain types of phones. For example, you might have standardized Google Android as your smartphone operating system platform, so you want to ensure that only Google Android devices connect to your corporate network. In other cases, you might want to scan for a particular version of an operating system or device type. For example, you might know that version X.2 of a vendor's operating system has some critical vulnerabilities, so you want to ensure that no device running that particular version of the operating system connects to the network.

Device type scans also help you determine any additional scans that you might want to run against a particular device. For example, you might have chosen a different antivirus/antimalware application for Android devices than for Symbian devices. Knowing the device type up front allows you to scan for the appropriate antivirus application when the device attempts to connect to the network.

✔ **Antivirus:** As discussed throughout this book, viruses, malware, and other types of exploits against smartphones are on the rise. Industry best practices for mobile device access are quickly settling on antivirus as a required application for the smartphone. Therefore, the ability to scan to ensure that an antivirus application is not only installed on the device, but also running and up to date, is becoming a key feature for many VPNs that provide endpoint integrity scanning.

Most SSL VPN vendors offer a solution that checks not only whether the machine has an antivirus application installed but also whether it's running and up to date. Some of the available policies on the market include

• Verifying installation of a particular version or vendor of antivirus solution(s).

• Verifying that real-time protection is actively enabled on the system.

• Verifying that virus signatures are fully up to date or that they've been updated at some point in the recent past, depending on your policy.

• Ensuring that a successful full-system scan has been completed in the last few days. (The number of days depends on your antivirus vendor's update schedule and your organization's willingness to allow machines with slightly outdated antivirus policies to join the network.)

✔ **Personal firewall:** This type of scan is fairly self-explanatory. Simply put, it determines whether a personal firewall is installed and running on the endpoint device. As firewalls increase in popularity for mobile devices, additional VPN vendors will begin to offer the capability to check for these important pieces of security software.

✔ **Disk encryption:** This functionality helps you determine whether encryption is enabled on the endpoint device. If you're familiar with encryption on traditional laptop and desktop machines, note that in the case of mobile devices, many of the device vendors have provided native encryption capabilities on the devices themselves, alleviating the need for third-party encryption products. In most cases, these encryption policies allow you to scan for whether encryption is enabled both on the embedded device disk and on removable media, such as SIM cards.

✔ **Antispyware:** You want to ensure that the antispyware application is not only installed, but also running and actively protecting the system.

✔ **Bluetooth:** Because a number of device exploits take advantage of Bluetooth capabilities on mobile devices (see Chapter 2), the ability to determine whether Bluetooth is enabled is important for some organizations.

✔ **Device lock:** This type of scan allows you to determine whether the appropriate idle timeout and lock policies are enabled on the device.

✔ **SIM policies:** You enable this type of policy to check whether the SIM card is PIN protected, and whether it is locked to the phone itself, helping to guard against theft.

An example of the type of mobile device integrity policy that you might see enabled on an SSL VPN gateway in a typical enterprise network is shown in Table 7-1. Note that this is an example, not necessarily all-inclusive or representative of best practices across every area.

Table 7-1	Example Mobile Device Integrity Policy
Attribute	*Allowed Values*
Device type	Apple iOS 4.0 and 4.1, 4.2, and 4.3 Google Android 2.0, 2.1, 2.2, 3.0, and 3.1 Blackberry OS 5.0 and 6.0 Windows Mobile 6.5 Windows Phone 7.0
Antivirus	Junos Pulse 2.x F-Secure Mobile Anti-Virus 2.x and v3.x Must be installed and running.
Encryption	Must be enabled.
Personal firewall	Must be installed and running.

Providing access based on device profile

Over time, many organizations have built access policies for traditional laptop and desktop platforms based on whether devices are known and/or managed:

- ✔ A **known device** is typically defined as a device that belongs to a particular user, and it is expected that when the user attempts to access the network, she will do so using the known device or devices associated with her profile.

- ✔ A **managed device** is also a known device, but typically, being managed means that the organization has more control over the device, and usually, it is a device that is owned and provisioned by the organization.

In the PC world, machine certificates have arisen as a very popular way to determine that a device belongs to the organization or falls into the managed category. Unfortunately, machine certificates have not yet made their way to most mobile platforms. Today, most organizations that wish to differentiate between known versus unknown and managed versus unmanaged mobile devices attempt to do so by verifying the *International Mobile Equipment Identity,* or IMEI number.

IMEI numbers are typically used by mobile operators to track devices and prevent use of stolen devices on the network. The number is unique to each device, so it can be leveraged to ensure that a device falls into a particular category before allowing access.

Note that the IMEI number identifies only the device, not the user, so in order to make use of this method of device identification, you must first ask your users to register their devices for enterprise use. This process typically involves the provisioning of the device into a corporate directory database so that the IMEI information can be retrieved at login time and associated with the user's profile to ensure a match.

Implementing custom policies

In some cases, organizations will have other things that they want to scan for on certain mobile devices, things that fall outside the capabilities provided by your VPN vendor. For example, you might want to scan to ensure that a particular application is installed on a device. Or you might want to ensure that an additional endpoint security application is installed and available. In the world of Microsoft Windows and Apple Mac laptops and desktops, this type of challenge is easily overcome. SSL VPN vendors offer a range of custom policies that allow you to search for files, processes, registry settings, and any number of additional attributes that allow you to identify these additional applications.

On smartphones, however, the challenge is much greater. A Microsoft Windows Mobile/Windows Phone smartphone is very similar to a normal Microsoft Windows machine from that perspective. Outside of that platform, however, your ability to check these various functions is difficult or even impossible. Many smartphones have implemented sandboxing functionality that severely restricts each application's access to the file system and to other applications running on the device. This effectively renders these custom checks impossible without some level of vendor support. Our prediction is that it will be quite some time before these capabilities that you already enjoy on Windows, Mac, and Linux systems make their way to the myriad smartphone platforms available in the market.

Providing Application Access

Before launching into a discussion on how VPNs can provide for granular access control to corporate networks, it's important to understand *what* you are providing access to. What are the types of applications that end users want or need to access on their smartphones? In our experience, many (though certainly not all) organizations have followed a similar curve in terms of application adoption for smartphones. Figure 7-2 illustrates how this works, and in this section, we discuss each of these types of applications and talk about their implications on security and VPN access.

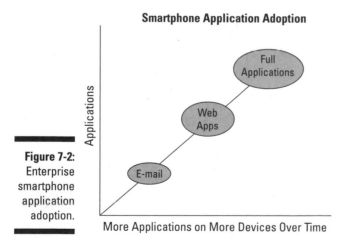

Figure 7-2: Enterprise smartphone application adoption.

Enabling access to e-mail

One of first things that many people do when they go out and buy that shiny new smartphone is try to figure out how to access their e-mail, calendar, and contacts. E-mail is a critical communications function in today's corporate world, and it seems that any time the e-mail system goes down, productivity plummets. Correspondingly, it simply makes sense that e-mail and messaging functions rank high in smartphone users' minds. If you can't provide end users access to their e-mail, they aren't likely to be very productive with their smartphones.

When today's smartphone platforms first started to become enormously popular, led by the popularity of the Apple iPhone, organizations started to bypass their own security policies in order to provide access to corporate e-mail. The story usually goes something like, "We never provided remote access for these smartphones until the day our CEO bought an iPhone and demanded access. Two months later, we checked our logs and determined that 2,000 people were now accessing our e-mail systems via smartphones using the same mechanisms that we provided to the CEO."

As smartphones have ushered in this era of end user choice, it's common to hear stories like that one. Unfortunately, many organizations have provided access to critical information — corporate e-mail — without proper regard for how to do so securely. Due to a lack of vendor solutions for this problem, and in many cases a lack of funds, access has been provided in a haphazard fashion.

Many enterprises require use of a VPN, strong authentication (such as a one-time password or a digital certificate), and a properly patched device in order to provide access to a managed laptop such as a Windows machine. Those same organizations provide access via smartphones with no VPN (by allowing their mail servers to be accessed directly), no strong authentication (and sometimes from devices with no timeout or lock password!), and no idea whether the end device is secured. The same security policies that they've spent years developing and refining have been thrown out the window in order to provide access for smartphones.

At the end of this chapter, in the section "Securely accessing e-mail, calendar, and contacts," we discuss some ways that you can provide this access, but do so securely using VPNs.

Providing Web application access

As organizations get more comfortable with the idea of providing access for smartphones, they typically start to look for ways to increase their usefulness as productivity tools by providing access to a broader range of applications. The second major category of application that enterprises typically adopt and roll out is web-based applications. Today's smartphones have very powerful Internet browsers and offer an experience not terribly different from that offered on traditional laptops and desktops. At the same time, many website and web application vendors have started to provide customized versions of their web applications that are better suited to smaller mobile device form factors.

So what are these applications that enterprise end users are accessing? The answer is that it varies widely, but fair game are any of those web applications you've deployed in your network that are considered critical for remote access. For some organizations, this means access to the intranet, or perhaps human resources and people-related applications related to expense reports and paychecks. For other organizations, this might mean access to mission-critical information such as web-based customer relationship management (CRM) tools and product or service information. Whatever the case, we describe how to provide access to these web-based applications in the section "Accessing web-based applications," later in this chapter.

Accessing full client/server applications

At the end of the application adoption curve is access to full, installed applications on the smartphone. These differ from web-based applications in that these applications are not accessed directly from a web browser, but rather, they are installed directly on the phone itself. In some cases, these applications are retrieved directly from the operating system vendor's application store, but as of this writing, these vendors are beginning to give enterprises tools that they can use to author and deploy their own applications from enterprise app stores.

Regardless of how the application is deployed to the device, these applications are important because they can currently provide a richer end-user experience than most web-based applications can provide. In addition, they are convenient. A simple button on the smartphone home page launches the application. As with web-based applications, the types of full applications leveraged by organizations run the gamut from fully installed CRM applications to applications that allow doctors to remotely view X-rays and patient medical records, for example. Across all of these types of applications, one commonality is that enterprises want to ensure that the data access through them is provided securely, and in the section "Allowing users to leverage client/server applications," later in this chapter, we show you how to do exactly that.

Providing Users an Appropriate Level of Access

After you know what end users will want to access, it's important to think about how the various levels of access can be provided. This section discusses e-mail access, web-application access, and full application access, and how the two most popular types of VPNs — IPsec and SSL — can help you meet those needs.

Securely accessing e-mail, calendar, and contacts

As discussed earlier, e-mail, calendar, and contacts are among the first applications that end users wish to access from their smartphones. Every major modern smartphone platform supports a set of protocols known as Exchange ActiveSync, a proprietary Microsoft protocol that allows for this same mail, calendar, and contact data to be transmitted between mobile device clients and a mail server. In many cases, that mail server happens to be a Microsoft Exchange server, but several other mail servers also support the Exchange ActiveSync protocol.

Connecting directly to Exchange Server via ActiveSync

While we always recommend use of a VPN, here are a few tips to ensure that your deployment is as secure as possible when connecting directly to Exchange Server.

Remember: Always use SSL encryption (and authentication) for connections between the mail server and the mobile device. You should never allow sensitive corporate data to transit on the Internet unencrypted. If you do not encrypt the connection, your organization's e-mail will transit the Internet in clear text. If that data is intercepted somehow, such as across an insecure Wi-Fi link that your user has connected to, it can easily be read by the intercepting person.

The Exchange ActiveSync protocols include a number of device security features that you can use to ensure that the data on each device is protected against loss and theft. All of these are best practices and should be implemented in accordance with your existing corporate security policies:

✔ Password complexity policies can be set on the mobile device remotely when it connects

to the mail server. For example, you can specify that a password have a minimum length and include a minimum number of alphanumeric characters.

✔ You can set a device lock timer, meaning that once the device has been idle for a certain amount of time, the device will automatically lock, forcing the user to enter the password again in order to access it.

✔ You can ensure that encryption on the device hard disk and removable media is enabled, ensuring that a phone that gets into the wrong hands does not contain easily readable data.

✔ You can remotely wipe a device, either automatically, such as when there have been too many failed password attempts, or by administrative command if the device is lost or stolen.

Ensure that your Microsoft Exchange Server (or other e-mail server) is always properly patched and up to date.

In order to provide access, many enterprises have simply deployed their mail servers so that they are externally reachable from the Internet, with the devices connecting directly to the server. There are advantages and disadvantages to using this approach. One major downside is that deploying the mail server on the DMZ exposes a very important asset — your corporate e-mail — as a target to the Internet. For this reason, we recommend, as a best practice, that organizations use a dedicated VPN for even basic e-mail access. On the other hand, there is no need to deploy any software on the mobile device in order to make this work because the major smartphone operating systems already support the ActiveSync protocols.

This section describes the use of a VPN as a proxy for Exchange ActiveSync traffic. If you have already decided to allow mobile devices to connect directly with the Exchange Server, however, check out the nearby sidebar, "Connecting directly to Exchange Server via ActiveSync."

Some vendor SSL VPNs allow your organization to proxy ActiveSync traffic without deploying any software onto the endpoint device. From a feature and user-experience perspective, this approach is no different from an approach where the end user connects directly to the mail server. On the other hand, the following benefits are associated with taking this proxy approach:

- **A VPN gateway is purpose-built to be hardened and secured.** These types of devices are specifically built and designed to be accessible from the Internet. As such, they have typically gone through numerous security audits, are regularly patched and updated, and have built-in protections against attacks that are generally faced by Internet-facing devices.

- **VPN gateways support strong authentication.** If your organization operates like a lot of others, you want to use strong authentication, such as one-time passwords or X.509 digital certificates, to identify users connecting to your network. VPN gateways support this functionality natively, so there is no need to provide alternate authentication mechanisms for your mobile device deployment.

- **The VPN approach allows you to standardize on a single platform for all of your remote access needs.** Because you are likely already using this type of gateway for access from traditional devices, you can ensure that all remote access into your network leverages a single termination point, simultaneously simplifying operations and reducing the number of devices you have exposed to the Internet.

- **Leveraging a VPN gateway today allows you to expand your scope as you support additional mobile device applications.** Because these devices support the ability to provide access to several different types of applications — as your mobile device deployment grows in size and becomes more strategic — you can include additional applications with the initial VPN gateway without swapping out or providing additional termination points in the future.

Accessing web-based applications

There are a number of ways to provide secure access to web-based applications, but for remote access to enterprise applications, one of the most common methods in use today is SSL, typically through an SSL VPN gateway.

Many web-based applications have built-in support for SSL termination and user authentication, but the problem that this chapter addresses is access to several applications, as in a typical enterprise intranet type of scenario. In this case, an organization could go through the expense of hardening and securing each application, providing authentication mechanisms and building an Internet-facing presence for each application, or the organization could use an SSL VPN to achieve this goal.

In fact, SSL VPNs were first brought to market for this very purpose, consolidating multiple web-based applications into a single Internet-facing portal. (They have since evolved to solve many different remote-access tasks within an organization, as described elsewhere in this chapter.) As an ever-increasing number of applications were moving to the web, the task of preparing each application for access from the Internet was increasing operational costs. SSL VPNs provided a way to simplify and consolidate. At the same time, they provided a way to provide access to third parties (partners and customers, primarily) without leveraging a full Layer 3 IPsec VPN connection onto the network.

This web-based mode of operation is sometimes referred to as a *clientless VPN,* acknowledging the fact that no client software needs to be installed on the endpoint device. Clientless SSL VPN functionality leverages only a web browser on the endpoint device, making it a ubiquitously available application, not only for traditional platforms, but also for a wide range of mobile devices. Clientless modes of operation on SSL VPNs remain a widely used deployment, largely due to these two key benefits:

✔ **No software is required on the endpoint device.** This simple fact makes SSL VPNs a perfect choice for access from any device. An end user can use an SSL VPN to access corporate data from his home machine, a kiosk, a mobile device, or really any machine with a web browser that supports SSL. As an added benefit, vendors have recently begun adding a range of features that allow the SSL VPN to optimize web-based application access for the smaller screen sizes typically found on mobile devices. This doesn't make access from these devices more secure, but the resized web applications and websites definitely improve the user experience.

✔ **Clientless SSL VPN solutions provide very granular control over end user access.** Because many organizations are only just beginning to embrace the use of mobile devices, and because many of them have yet to roll out some of the security and protection mechanisms described throughout the rest of this book, providing tight control over what a user can access is an attractive value proposition.

In many clientless SSL VPN implementations, web-based application access can be controlled all the way down to the individual file or URL level. So if a remote user should have access to only one particular file or application, leveraging an SSL VPN can ensure that the remote user can't see or access any other applications in the corporate network.

How does the clientless mode of operation work? It depends on the implementation, and most vendors have developed this key intellectual property over time. For the most part, clientless SSL VPNs use something called a *rewriter*, which actually intermediates every request and response that goes through the SSL VPN, and modifies embedded links so that, to the outside world, the content appears to be served directly from the SSL VPN.

This rewriting capability provides granular access control and, at the same time, allows organizations to mask the details of their internal application deployments from would-be hackers. If a hacker can easily get the IP address or URL of an application server that's housed inside the network, he or she can begin to formulate a plan for attacking that server, a less-than-desirable outcome for your network.

One of the downsides of a clientless SSL VPN is that this type of access method does not allow users to access fully installed, client-server applications. They really can handle only web-based application traffic. For providing secure remote access to client-server applications, a client application that provides a tunnel into the corporate network is required. We show you those solutions in the next section.

Allowing users to leverage client/server applications

As described in the earlier section, "Accessing full client/server applications," the number of organizations providing access to fully installed client-server applications from mobile devices is growing with every passing day. Despite predictions over the years that eventually all applications will be web-based applications, those days are still a long way out (if they ever arrive), so a solution is required to provide a secure tunneling mechanism for data from these applications destined for the corporate network.

In this section, we discuss three main types of VPN clients that accomplish this task of tunneling client-server applications. With many of these technologies, vendors have developed client-based technologies that frequently combine some of the advantages of clientless SSL VPNs with some of the access required for these types of applications. Here are the three main types of clients that we discuss:

- ✔ IPsec VPN Layer 3 network-extension clients
- ✔ SSL VPN Layer 3 network-extension clients
- ✔ SSL VPN Port forwarding client applications

Although most SSL VPNs and some IPsec VPNs do offer dynamic delivery and installation of client software, there will be limitations on dynamic delivery depending on the smartphone platform. For example, as of Apple iOS 4.3 (currently shipping at press time), there is no way to dynamically provision software to an iPhone. As a result, the process requires that end users first visit the Apple App Store and download the VPN client. From there, they need to point the client to the appropriate VPN gateway, typically by inputting a URL or address. Users then provide their credentials, and the VPN connects. This isn't a terrible end user experience, but it's definitely a far cry from the dynamic deployment methodologies that many organizations have gotten used to leveraging over the past few years.

Using IPSec VPN clients to provide full Layer 3 connectivity

IPsec VPN clients provide a full, Layer 3 connection to the corporate network. Up until the mid-2000s, IPsec VPNs were the primary method of remote access. Today, many mobile devices include an embedded IPsec VPN client, providing a bit of a renaissance for remote access IPsec VPNs. (They have remained popular for site-to-site VPNs since their inception.)

Regardless of whether the IPsec client is embedded or a third-party application is installed on the mobile device, IPsec VPNs provide the LAN-like connectivity required to support the broadest possible range of applications on a mobile device. From web-based applications to very advanced multimedia applications such as Voice over IP and video, IPsec VPNs provide a mechanism by which an organization can securely tunnel some or all of the data from a mobile device through the corporate network.

IPsec VPNs have become less popular for remote access because deployment of client software has historically been a challenge, and because IPsec VPNs provide only one type of access: full Layer 3 connectivity into the network.

In many cases, this level of connectivity is simply overkill and exposes more information than is required to the end users. It is always a best practice to limit access only to that which is absolutely required for a given user or group of users.

Over the last couple of years, several IPsec VPN vendors have begun to add technology that allows for easier deployment of client software, using mechanisms traditionally employed with SSL VPNs. Additionally, as mentioned before, a number of modern smartphone platforms offer embedded IPsec VPN support, obviating the need for a client. It's always a good idea to take a survey of both your short-term and long-term requirements and choose a VPN technology that meets all of those needs, across all the client platforms that you anticipate allowing into your network.

Leveraging SSL VPN Layer 3 network extension clients

Layer 3 clients, offered by every SSL VPN on the market, are very similar to traditional IPsec VPNs in terms of the connectivity that they offer. When a user is granted access through one of these clients, he or she is assigned an IP address and has full, LAN-like network connectivity.

Connecting through a full Layer 3 SSL VPN client gives the user the same type of experience that he or she has when connecting through an IPsec VPN or attaching directly to the LAN itself.

The SSL VPN version of a Layer 3 client offers some significant advantages over traditional IPsec VPNs. Among the key advantages is the much easier deployment of an SSL VPN client. Installing and configuring IPsec VPNs can be difficult, typically requiring manual intervention to set some of the configuration options. With SSL VPN, however, most offerings provide dynamic delivery of the client. When the user first needs to log in to the SSL VPN appliance, he or she browses to the appropriate URL.

After the authentication and authorization process has been completed by the SSL VPN, if the policy states that the user should have full Layer 3 access, the client is dynamically delivered and installed on the user's machine with no intervention required on the administrator's part. The SSL VPN appliance handles upgrades in a similar fashion, seamless to both the administrator and the end user.

When evaluating mobile device VPN solutions, make sure to pay attention to platform support. Especially during the early days of your mobile device deployment, you might have opinions on certain platforms that will change over time. For example, you might think that you can restrict access to only Apple iPhone and Google Android platforms, so you select a VPN that provides VPN clients for these devices only. Then, along comes a new platform that rockets to the top in terms of popularity. If your vendor doesn't respond to provide access to this platform, you're going to be stuck deploying another VPN.

Ensure not only that your vendor provides adequate support for current-generation platforms, but also that the vendor is committed to staying ahead of the curve with new devices and platforms as they're introduced to the market.

Using SSL VPN and port-forwarding client applications

Not all SSL VPNs provide port-forwarding technology, but you may see it. Like with the Layer 3 network extension, the *port forwarder* is a dynamically installed and delivered client application that provides access to full versions of client-server applications.

The primary difference between port-forwarding applications and Layer 3 network extenders is that the port forwarder controls access at a more granular level, specifying exactly which resources can access the VPN. With these technologies, the application believes that the client application is the destination application server. The client then intercepts this traffic and forwards it to the SSL VPN appliance over the secure connection, where it's then forwarded to the final destination, the application server.

 In some cases, the port-forwarding application can also specify which processes on the client side are allowed to access the tunnel, not only the destination. In other words, you might have a policy that states that only Mobile Outlook can use the connection, and it can only pass traffic to certain ports on the Microsoft Exchange messaging server. You get a much more granular level of control than a Layer 3 network extender can provide, and depending on your organizational needs, you might use port forwarding for some users and network extension for others. This is a policy choice rather than a hard-and-fast rule. For example, we frequently see organizations provide full network extension for employees, but port forwarding for only a defined set of applications for partners, specifically because their security policies prohibit partners from having full network access.

Note that port-forwarding applications have yet to really take off in the mobile device market, so very few vendors provide this type of solution for smartphone platforms. It is likely that you might find this solution for one or two smartphone platforms, but not for all. A Layer 3 VPN — either SSL or IPsec VPN — will likely be the only choice that you have for access across the wide range of smartphones that your organization might want to allow onto the network.

Case Study: AcmeGizmo SSL VPN Rollout for Smartphones

Returning to our ongoing case study on AcmeGizmo, Ivan, the IT manager, is at the point where he wants to start exploring how to securely connect employee mobile devices to the corporate network and protect sensitive data as it transits the Internet.

Recall from Chapter 1 that AcmeGizmo currently offers three mechanisms for connecting into corporate from any device. Users on corporate-issued BlackBerry devices connect to the network via AcmeGizmo's BlackBerry Enterprise Server. Users on corporate-issued Windows laptops connect to the network via an IPsec VPN solution from a company called Connect PC. Finally, Ivan recently began to allow a select group of mobile devices to connect directly to the corporate mail server, though he has discovered that the word has gotten around and, to his surprise, over 500 devices are now connecting via this mechanism. Figure 7-3 shows the current remote access strategy at AcmeGizmo.

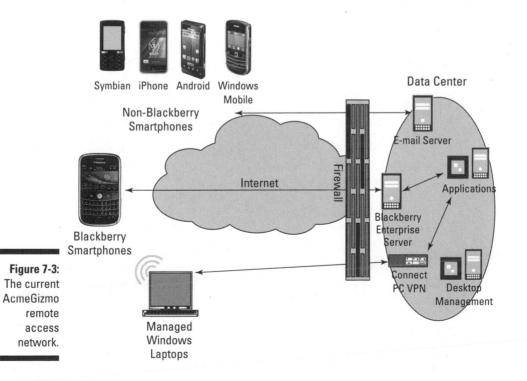

Figure 7-3: The current AcmeGizmo remote access network.

While in the process of securing his mobile device remote access deployment, Ivan would like to consolidate remote access to fewer appliances in the corporate data center. His rationale is that this will help reduce management and administration overhead, while simultaneously lessening the probability that a misconfiguration or vulnerability will result in data theft.

Three primary groups of employees (a general group, executives, and salespeople) need Ivan to provide them with mobile device access to the network. Each requested access to different sets of applications and data. Ivan's challenge is to put into place a strategy that serves all of these employees across the different sets of devices that they wish to bring into the network.

Employee authentication

Ever since he found an employee's password affixed via a sticky note to the bottom of an AcmeGizmo laptop, Ivan has been considering a stronger authentication solution than the standard username and password requirement that AcmeGizmo has been using for years. He has decided that he will employ a stronger authentication solution for all employees as he makes the transition to the new remote access strategy.

The new policy will be that all users accessing the AcmeGizmo corporate network from outside the local area network (LAN) must use the two-factor authentication solution that he has purchased. Based on preference, some employees will carry with them a hardware token that updates automatically every 60 seconds with the newest one-time password. Other employees will not need to carry a token, but instead will be sent a text message when they attempt to log in to the corporate network remotely. That text message includes the one-time password.

Each employee enters his username, personal identification number (PIN), and the current token in order to authenticate to the network from any device. When the employee is authenticated, the new remote access system also queries AcmeGizmo's Active Directory (AD) server in order to determine which group(s) are assigned to that employee (Executives, Enterprise Sales, or the general Employees group). This helps the remote access system determine the level and type of access each specific user will get.

Accessing the network with SSL VPN

Ivan took a look at the various VPN offerings available and decided to purchase an SSL VPN from Juniper Networks since it integrates with the endpoint security solution for mobile devices that he is also looking at from the same company.

Ivan decided that all remote access users from Windows laptops and Apple iOS, Google Android, Nokia Symbian, and Windows Phone devices will access the corporate network through the SSL VPN. The only device type in the AcmeGizmo network that will not leverage the SSL VPN is the existing BlackBerry devices. Ivan recognizes that the BlackBerry Enterprise Server (BES) that's already deployed in the network is a feature-rich and secure single-platform solution. Rather than remove the BES and migrate all the existing users, he has decided he will continue to leverage the deployment that is up and running. Ivan also has to ban any device types other than the aforementioned devices from the corporate network.

On the device itself, all employees will be asked to download the Junos Pulse client software. In some cases, this is deployed to the device via a text message sent to the device upon registration. In other cases, the employee downloads the software from the application store for their particular device. Regardless, this endpoint software is the single agent that Ivan needs to ensure resides on each device.

When attempting to connect to the network via Junos Pulse, each user is prompted for his username, PIN, and one-time password, as described previously. From there, the SSL VPN groups that user into one of three roles and assigns them the appropriate level of access.

General employee access

In the past, most employees were not provided with a corporate-issued BlackBerry, so they didn't have access to their e-mail, calendar, and contacts when not in front of their laptops. Ivan's boss, Steve, however, feels strongly that allowing this level of access could boost productivity across the company, so he has informed Ivan that he would like to figure out a strategy that will allow all employees to access their e-mail, calendar, and contacts, but restrict access to everything else from their mobile devices. These employees don't need to download and install Junos Pulse on their devices. Instead, they leverage the native configuration in their smartphones to have them connect directly to the SSL VPN appliance via the ActiveSync proxy functionality, which will secure the connection and provide a very limited access for these employees.

Executive access

The executives within AcmeGizmo who wish to access corporate data require access to a wide range of applications. E-mail, calendar, and contacts are, of course, critical for each of them to have access to. Beyond that, they require access to several sites on the AcmeGizmo intranet, as well as access to several applications that have been purpose-built for smartphones, including the company's customer relationship management (CRM) application.

As with the other groups of employees, each executive has Junos Pulse installed on his or her machine. When the executive logs into Junos Pulse, the software determines that the person in question is in the Executives group and provisions a full Layer 3 SSL VPN tunnel into the corporate network. This ensures that each executive has access to all aspects of the AcmeGizmo network. In fact, the full Layer 3 access provides an experience similar to the experience that the executive has when accessing the corporate network from his or her Windows laptop on the corporate LAN.

Enterprise Sales access

The Enterprise Sales team has only a very specific access need above and beyond what the general employee group requires. These employees require access to the intranet and the company's CRM application, both of which are web-based applications based through the company intranet. As with the executives, these employees have Junos Pulse installed on their mobile devices. When they log in using their one-time passwords, the SSL VPN system identifies them as a Sales user, automatically provisioning access to their e-mail, calendar, and contacts; but it also provisions bookmarks in Junos Pulse, pointing them to both the intranet and the CRM site by leveraging the rewriter function in the SSL VPN.

Figure 7-4 shows the new AcmeGizmo remote access deployment. As you can see, in addition to making strong authentication a mandatory part of access into the AcmeGizmo network, Ivan has been able to consolidate the number of entry points into the network, and also remove the e-mail server from the demilitarized zone (DMZ).

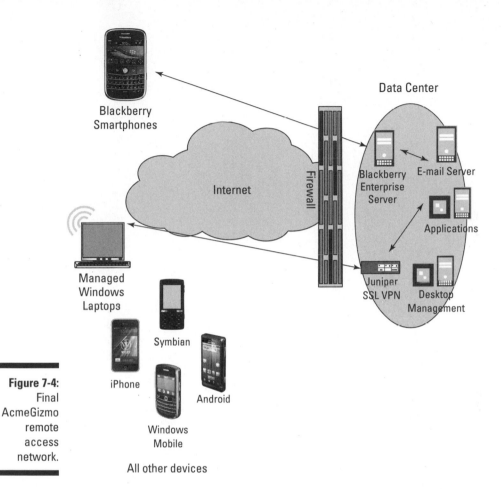

Chapter 8

Connecting to Wi-Fi Networks

Mobile devices such as iPhones, iPads, and those running the Android operating system have sophisticated Wi-Fi capabilities, allowing them to connect to public and private networks for Internet access. Device users can connect to networks at public places, such as coffee shops, airports, and hotels, and to private networks, including corporate and home networks.

This chapter explores the world of Wi-Fi on mobile devices and describes ways users can connect to networks and how you can manage policies and settings for Wi-Fi access. We also discuss the risks of users connecting to certain public Wi-Fi networks, especially those that are open and allow any device or user to connect to them. Finally, we look at options for securing your corporate Wi-Fi network.

What's Wi-Fi, and Why Bother?

Smartphones today have access to the wireless carrier's data network, enabling them to send and receive data such as e-mails and text messages. Wireless carriers have built elaborate networks to handle the load of millions of smartphone users. In many cases, these are 3G (or third-generation) networks, and some carriers have even built more sophisticated 4G or LTE (Long Term Evolution) networks. 4G or LTE networks have greater capacity and bandwidth than the older third-generation networks.

In many cases, however, such networks have inadequate strength, causing devices to either lose network coverage or experience slow network access. Most people have experienced network outages of this sort, especially in crowded cities or downtown locations where many devices compete with one another for access to the carrier's network.

Enter *Wi-Fi* technology, which is designed to connect computers or other devices within short distances without needing cables. Wi-Fi allows the connection of multiple devices into a single network, all of which can then browse the web, send e-mail, and connect to the Internet. In your organization, you can create a corporate Wi-Fi network to which employees connect their various devices, including laptop computers, smartphones, and tablets.

Wi-Fi networks provide a sigh of relief to smartphone users in counteracting the unpredictability of carrier networks. Wi-Fi networks provide Internet access in various locations, such as hotels, airports, and coffee shops. Users at these locations can get off their 3G networks and connect to a typically more stable, and often faster, Wi-Fi network. Many Wi-Fi networks are public or insecure, meaning that any device or user can connect to them. Insecure networks come with some risk, exposing users to the possibility that their data (such as e-mail or web pages) could be read by other people who are connected to the same network.

With the increasing number of smartphones and people using these devices for browsing the web or sending e-mails, the appetite for network capacity is increasing rapidly. Therefore, Wi-Fi networks are being deployed at more and more locations, providing network access to millions of users every day.

Which Wi-Fi Networks Should Users Connect To?

Not all Wi-Fi networks are secure. Some are *open* networks, requiring no authentication of the devices or the users connecting to them. These open networks may be deployed at airports or coffee shops. All it takes for a user to connect to such a network is to detect the open network by name (or Service Set Identified [SSID]) and connect to it. No password is required, thereby letting anyone connect to the network.

Wi-Fi networks can be secured by requiring a password or using other techniques. Such networks are relatively more secure to connect to. But depending on the nature of security deployed in the Wi-Fi policy, these networks can also be snooped on. In the following sections, we look at the two broad categories of Wi-Fi networks to which users can connect their smartphones and tablets.

Open or insecure networks

Open networks can be joined by any user and from any device without the user needing to enter a password. These networks are typically the riskiest for users to connect to, because the data transmitted and received by users can be viewed by other users connected to the same network. People conversant with networking technology can read traffic over the network sent by other users from laptop computers, tablets, or smartphones.

Traffic that is easy to snoop on includes open or unsecured browsing traffic, such as visiting a website that does not require SSL encryption. Unfortunately, many popular websites like Facebook, Yahoo!, and Twitter do not need SSL encryption, so when users browse these sites over an open Wi-Fi network, they're vulnerable to being snooped on.

Websites or applications that require SSL encryption are more secure from being snooped on by users on the same Wi-Fi network. While browsing to any website, users can easily see if SSL encryption is turned on. It usually appears as a padlock on the browser itself, sometimes along with the name of the server the user is connecting to. When users browse to websites that do not need SSL encryption, their information is visible and readable by others if they are on an open Wi-Fi network.

If you're managing mobility policies for your corporate users, you need to strongly discourage them from connecting to open Wi-Fi networks from their smartphones or tablets. Whether your employees are using personal devices or corporate-owned devices, you don't want users on an open Wi-Fi network.

Encrypted Wi-Fi networks

Wi-Fi networks can be secured using techniques called WEP (Wired Equivalent Privacy), WPA (Wi-Fi Protected Access), or WPA2 (a more recent form of WPA). Among these three, WEP employs the weakest encryption, because it relies on a preshared password key, which is used to encrypt network traffic. WEP-secured networks are more secure than open networks, but anyone who has successfully connected to a WEP-encrypted network can view traffic generated by other users or devices on the same network.

WPA and WPA2 employ stronger encryption than WEP. WPA2 uses stronger encryption and is more recent than WPA. WPA2 comes in two flavors: WPA2-enterprise and WPA2-personal. For private networks, such as home networks, WPA2-personal is the ideal security to deploy. For corporate Wi-Fi networks, WPA2-enterprise is the best possible security to deploy.

As an administrator recommending mobility policies, you can feel secure if users are connecting to WPA2-secured Wi-Fi networks from their smartphones and tablets.

VPN on a Wi-Fi network

If your users do happen to connect to open an Wi-Fi network, make sure they use VPN on their devices to connect to your corporate VPN gateway. VPN results in a secure tunnel being built from the device to the VPN gateway, through which all traffic is encrypted and invisible to network snoopers. VPN comes in IPSec and SSL flavors, both of which have their pros and cons. Most laptop PCs, Apple Macs, smartphones, and tablets include VPN support for leading networking vendors.

Chapter 7 describes VPN in greater detail and runs down the corporate options to use VPN on various devices.

As previously mentioned, you want to strongly discourage users from connecting to open Wi-Fi networks. However, your users may connect their devices to open networks anyway, so you must consider VPN as a required policy in such cases.

Wi-Fi Connections from Mobile Devices

In this section, we look at how users can connect to Wi-Fi networks from their mobile devices. We focus on the most popular types of devices available: Apple iOS, Google Android, and BlackBerry. We run down some techniques you may want to use when instructing your corporate users. We include signs that users need to watch out for, such as the absence of security on a Wi-Fi network. Make sure that users are wary and cautious of connecting to open networks.

Apple iPhones, iPads, and iPods

The Apple iPhones, iPads, and iPods all run a single operating system, called the Apple iOS, which makes the configuration of Wi-Fi identical on each of them.

Here are the steps that your users need take to connect to a Wi-Fi network — at home, at a public location, or at work — from devices running the Apple iOS operating system:

1. **Navigate to Settings⇨Wi-Fi on the iOS device. Make sure Wi-Fi is On.**

   If you don't see any networks listed, that means you and your device aren't close to any Wi-Fi networks. If there are Wi-Fi networks in your vicinity, those should be displayed on this page, as shown in Figure 8-1.

Figure 8-1: Browsing Wi-Fi networks on an iPhone.

2. **Tap the network that you want to connect to.**

   If the network is open or public, you don't need to enter a password; you should be able to connect right after you tap the network.

   If the network is secure, you're prompted to enter the password to connect.

   If you don't see a padlock symbol displayed next to the network, that network is open or insecure. You can click the blue arrow at the end of each row to find more information about the network and the encryption technique used.

3. **If prompted, enter the password to connect to the network.**

   In most cases, the preceding steps should connect you to the nearest Wi-Fi network. If, however, you're trying to connect to a hidden network that isn't displayed on the device, tap Other in Wi-Fi Networks Settings to manually enter the network's information on the screen shown in Figure 8-2.

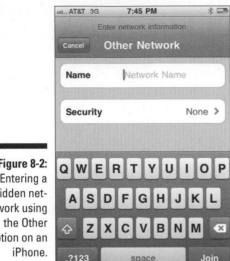

Figure 8-2: Entering a hidden network using the Other option on an iPhone.

You need decent signal strength to connect to a Wi-Fi network. If the network isn't close by or the signal isn't strong enough, you may not be able to connect to it.

Many public networks require you to accept a legal agreement to use them. For example, if you connect to a Wi-Fi network in a hotel, more than likely a web page will be displayed with legal disclaimers, asking you to accept or decline the agreement. The disclaimer usually indicates that you choose to use the network at your own risk, and that the network owner isn't liable for damages or losses that you may incur on their network. Unless you accept the agreement, the network will prevent you from browsing the Internet.

Connecting to Wi-Fi with Android devices

A number of devices on the market run the Android operating system, including the Motorola Droid, HTC Desire, and Droid Incredible.

Here are the steps that your users need to follow to connect to a Wi-Fi network from devices running the Android operating system:

1. **On your Android device, tap the Settings icon.**

2. **Tap Wireless & Networks under Settings.**

   The Wireless & Network Settings screen appears, as shown in Figure 8-3.

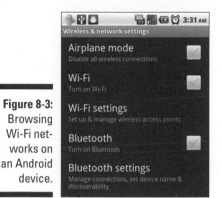

Figure 8-3:
Browsing
Wi-Fi net-
works on
an Android
device.

3. **Tap Wi-Fi Settings.**

   Note that Wi-Fi needs to be turned on for the device to be able to detect Wi-Fi networks. If necessary, tap Wi-Fi to turn it on.

   When Wi-Fi is enabled, you should see a list of networks on this page. (If you don't see any Wi-Fi networks, it simply means that you aren't near any.)

   Networks that are open and insecure appear without a padlock icon. Beware of connecting to such networks. At the very least, start up a VPN tunnel to your corporate VPN gateway as soon as you connect to such an open network.

4. **Select a network to connect to.**

5. **If necessary, enter the password to connect.**

   Many public networks require you to accept a legal agreement to use their network. Be sure to glance through the disclaimer before you accept the policy offered to you.

BlackBerry devices

Users can easily configure their BlackBerry devices to connect to public and private Wi-Fi networks. Like Apple iOS and Google Android devices, BlackBerry devices can deliver the same data services over Wi-Fi as on the user's cellular network and potentially faster download speeds. That means users can access their e-mail and browse the Internet just like on the cellular network.

Some BlackBerry device models support UMA (Unlicensed Mobile Access), which is provided by some carriers around the world. UMA allows users to make phone calls over a Wi-Fi network, allowing their friends and contacts to reach them on the same mobile number anywhere around the world. As long users are connected to a Wi-Fi network, their phone calls can be routed over it, without needing to rely on a carrier network.

Here are the steps that your users need to follow to connect to a Wi-Fi network from devices running the BlackBerry operating system:

1. **Select the Manage Connections option on the main menu.**

2. **Select the Set Up Wi-Fi Network option.**

3. **Select the option to Scan for Networks.**

   The device should automatically detect Wi-Fi networks in the vicinity. If not, you need to manually enter the network's information.

4. **Select a network to connect to.**

5. **Enter a password if the network requires one.**

   If you connect to an open network, it won't prompt for a password.

Implementing Wi-Fi Policies

In most cases, once a smartphone has been used to connect to a particular Wi-Fi network, it remembers the network for future use. This means that whenever that network is in the vicinity of the device in the future, the device will connect automatically.

Private Wi-Fi networks, such as home networks, are best secured using WEP or WPA/WPA2 encryption. If your users are setting up a Wi-Fi network at home, they need to be sure to use these techniques to set up a suitably secured Wi-Fi environment.

If you're deploying a corporate Wi-Fi network for many users, you should be looking for an enterprise-grade Wi-Fi with WPA2-enterprise encryption. This form of encryption may require you to deploy other infrastructure servers, so be sure to investigate the options from your networking vendor.

For corporate Wi-Fi networks, you often need to provision policies and settings indicating the networks available in a corporate building. These policies include the name of the network and the password used to secure the network.

As an enterprise administrator managing policies for many users, you want to set policies that push out names and security keys of secure Wi-Fi networks that you want users to connect to, which may include your corporate Wi-Fi networks worldwide. When users bring their devices into the work environment, their devices will then detect and connect to the network, without needing the user's intervention. This setup is ideal because it forces users to be on the corporate Wi-Fi network whenever available. When users move out of reach of the Wi-Fi network, their devices fall back to the carrier network.

In the following list, we look at the choices available to deploy Wi-Fi policies to smartphones from an enterprise perspective:

✔ **iPhone and iPad:** An application called iPhone Configuration Utility, shown in Figure 8-4, enables you to configure policies to enforce on corporate users' iOS devices. These policies include Wi-Fi configuration as well. When you create a policy, the iPhone Configuration Utility produces a profile that can be sent out to all users at once. Users need to install the profile from their iOS devices to activate the policies and settings you've set up.

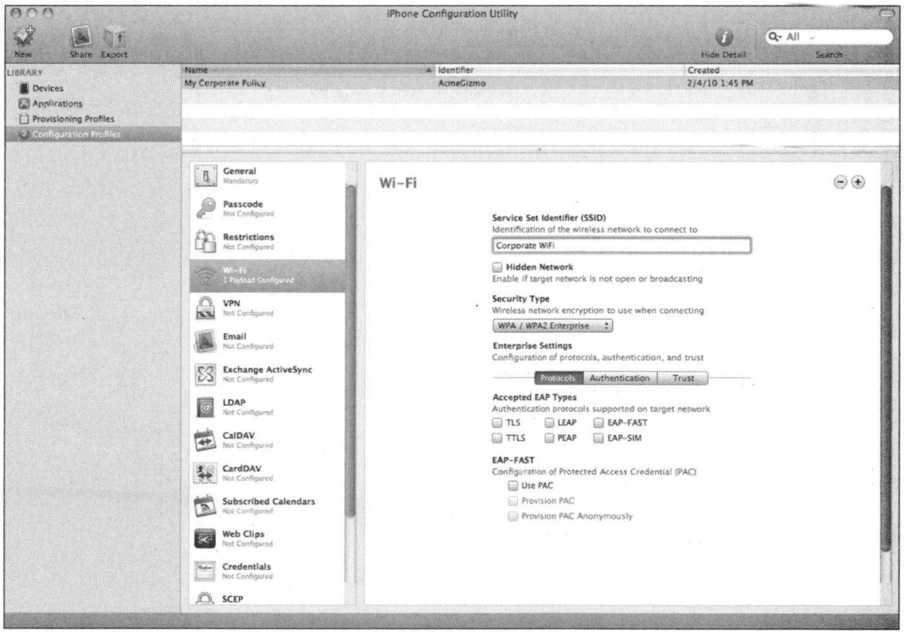

Figure 8-4:
The iPhone Configuration Utility allows configuration of Wi-Fi policies for iPhones and iPads.

Mobile Device Management (MDM) vendors offer the feature of deploying such policies to Apple iOS devices centrally. You can utilize an MDM solution to define Wi-Fi policies, passcode settings, and many other policies, and deploy them centrally to all iOS devices with one click. We discuss these solutions in more detail in Chapter 15.

✔ **BlackBerry:** The BlackBerry Enterprise Server manages the configuration and deployment of Wi-Fi policies across all BlackBerry devices used in an enterprise.

The BlackBerry Enterprise Server supports a variety of policies, including device encryption, passcode compliance, and browsing preferences. You can centrally administer these policies and deploy them to all BlackBerry devices at once.

✔ **Android and Windows Phone 7 smartphones:** Google and Microsoft provide no solutions to manage corporate Wi-Fi policies. If you need to configure policies for all types of mobile devices, including iPhones, iPads, and Android devices, look for Mobile Device Management (MDM) solutions, which are available from vendors such as Juniper, Good Technology, and MobileIron. For a detailed review of MDM solutions, be sure to read Chapter 15.

Part IV
Securing Each Smart Device

The 5th Wave   By Rich Tennant

"Russell! Do you remember last month when I told you to order 150 SMARTphones for the sales department?"

In this part . . .

It's time to roll out the policies, programs, and technologies to encrypt, protect, and back up the devices in your network. We walk down the thorny road together until you're comfortable being in charge of access and control.

Chapter 9 gives an overview of everything you can do. Chapter 10 tells you about device-based solutions, and Chapters 11, 12, and 13 suggest solutions you can implement both on the device (hard), in the brain of the device user (harder), or system-wide using the network tools in your arsenal (easy). AcmeGizmo is doing it, and doing it right, so catch the case study in the chapters.

You're just a few chapters away from reclaiming your network's integrity.

Chapter 9

Device Security Component Overview

▶ Identifying the various components of device security

▶ Protecting devices with on-device Anti-X protection

▶ Knowing in advance your backup and restore capabilities

▶ Incorporating loss or theft protection

▶ Controlling user behaviors (yeah, right)

▶ Managing devices in the enterprise

*T*his chapter introduces the various on-device security components that provide a fairly robust security envelope when used smartly. You (and your users) need to understand these various components to be well equipped to harness the capabilities of these features, making your collective lives easier and more manageable.

Knowing Smartphone Security Components

Each of this chapter's smartphone security components brings with it a unique and distinct capability that, when used wisely, provides you with ammunition to counter the various nefarious forces that are battling to gain access and compromise these devices.

We call to your attention to six discrete areas: device-based Anti-X protection, backup and restore capabilities, loss or theft protection, application control and monitoring, enforceable encryption, and enterprise management.

These six components are different enough in the type of protection that they provide that it's not just a question of whether to use one or the other, but how to use them all. This chapter helps you understand specifically what type of protection these components provide so you can turn and twist their dials and create something that makes sense to your organization. Not to mention what to implement first.

Note that this book uses the phrase *implement first,* which literally means that you need to prioritize the rollout of these components. Make no mistake, you need to eventually arrive at a complete security strategy when all are part of your security arsenal, but let's be real. You need a horse in front of the cart to get things rolling.

Consider these six components of device security (shown in Figure 9-1):

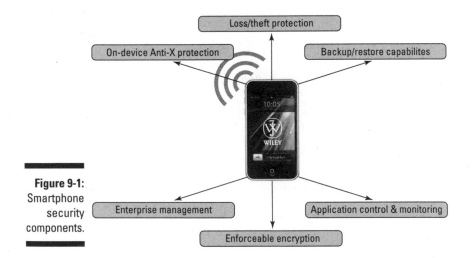

Figure 9-1:
Smartphone
security
components.

✓ **On-device Anti-X protection:** The security software actually running on the smartphone device itself

✓ **Backup and restore capabilities:** The ability to back up the information resident on the smartphone, including applications and data and their configurations

✓ **Loss or theft protection:** The remediation and recovery capabilities in the event of loss or theft of the smartphone itself

✓ **Application control and monitoring:** The enforcement of corporate policy as it relates to the usage of applications by users of smartphones

✓ **Enforceable encryption:** The ability to compel obfuscation of data — both resident on the smartphone as well as in-transit to the enterprise

✓ **Enterprise management:** The overall provisioning, troubleshooting, upgrade and monitoring of these smartphones

Understanding On-Device Anti-X Protection

When you are responsible for the device in the enterprise, this includes all of the associated applications, data, and the security posture of the smartphone or device. One of the key security components that is relevant to the security on the physical smartphone device is the "Anti-X" protection on the device. *Anti-X* refers to the family of security components that includes antispyware, antivirus, antiphishing, and antispam, as shown in Figure 9-2, and as the name suggests, can be extended to other threats that may arise in the future. So what exactly are these various subcomponents? Let's delve into each one. You're probably familiar what they are in terms of laptops and desktop computers, but mobility changes everything, including the equation that X equals security risk.

Figure 9-2: Smartphone security components.

Antispyware

In the term *antispyware*, the *anti-* refers to the essential component of the protection afforded against malicious spyware that installs itself on mobile devices. As a mobile device is always on the go — and with the plethora of interfaces supported by these smartphones — the likelihood that the smartphone is connected to one or more wireless networks most of the time is very high. This constant nomadic behavior and propensity to tethering means that the exposure level to unknown networks is very high, and therefore the likelihood of intrusions that can happen on these devices is far greater than a fixed desktop.

There are some unique dimensions to mobile spyware that make it different from the traditional desktop spyware that you might be used to. For instance, there have been cases of spyware that manipulate SMS messages and expose them so that they can be read by others in the near vicinity, as shown in Figure 9-3.

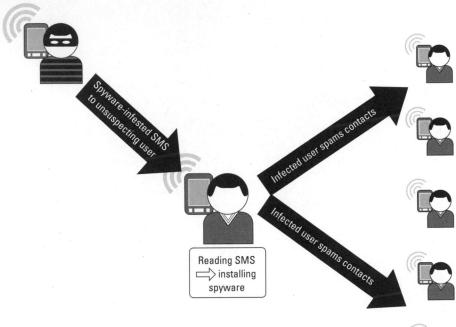

Reading SMS
⟹ installing
spyware

Figure 9-3:
Mobile
spyware in
operation.

In the figure, an unsuspecting user is tricked into reading an SMS message that has spyware associated with it. This could be as simple as a URL in the SMS that the user clicks, which lands him on a malware-infested website. In this instance, the spyware scrolls through the contact list on the mobile device and starts spamming the contacts using every means possible — SMS, e-mail, IM, and so on.

So any antispyware solution for mobile devices — in addition to protecting against traditional spyware, such as keyloggers, data leakage, *botnet membership* (membership in a group of infected devices that have been taken over surreptitiously by hackers), and so on — needs to provide specific protection against mobile threats to mobile applications (such as SMS-based spyware), contacts database protection, location information *spoofing* (masquerading the device location to be any place of choice), and the like. If you think that's still science fiction, think again. Do a Google search of the word *spy phone,* and the top hits you'll get undoubtedly include ads for spyphone software intending to turn innocent devices into recording devices that send the records of all activities to a designated place. Yikes.

The market is there for the asking, and that means that hackers will be coming after your users' devices in a big way, if not now, at some point in the near future. The simple solution: Be prepared to address the future with smart devices that have an antispyware solution.

Antivirus

Antivirus is a technology that has been available for decades, and many of your users would never consider operating a computer without some antivirus solution running on it. They get it when it comes to their desktop computers. However, a majority of mobile devices — which are all derivatives of computers in one way or another — go around without any sort of antivirus protection on them whatsoever! What is even more surprising is the despite this fact, your users (and you) increasingly rely more heavily on and become more personally attached to the smartphone. It's like wearing a sweater at home on a cold day, but ignoring your coat when you go outside.

You need to take a stand and ensure that you're providing adequate mobile antivirus coverage to your users on their mobile (and desktop) devices. The breadth of antivirus solutions is ever-increasing. Just as with traditional antivirus solutions, you should be looking for upfront costs; per-seat license renewals; automatic signature updates; and more uniquely mobile features, such as battery life recognition, memory requirements, and broadest mobile operating system coverage.

One tried-and-true antivirus solution comes from the traditional client-server model. In this scenario, an antivirus agent is downloaded to the device, but a bulk of the intensive processing that antivirus demands is actually performed on the server (either locally hosted by you or by a hosted cloud service). The client collects information about the mobile device and delivers a certificate of authority. In this model, shown in Figure 9-4, there may also be a *clone* (or virtual smartphone, as shown in the figure) of the actual enterprise phone maintained by you in the enterprise (maybe in the form of a virtual machine), and the agent informs you of any changes to the end device, such as new applications installed, SMSs received, and so on, and then syncs with the virtual phone in the enterprise.

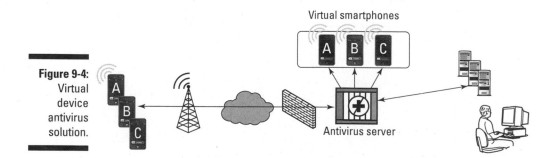

Figure 9-4: Virtual device antivirus solution.

Virtual smartphones

Antivirus server

This is not real-time protection of the device, but it's reasonably close and has the advantages of not causing performance or battery drain issues. In addition, because the antivirus solution is hosted on a server, there's a lot more horsepower than is available for antivirus checking on the device.

Antiphishing

Phishing attacks on mobile devices are likely to be far greater than they are on your standard laptops and desktops. The reasons for that, as follows, are fascinating to consider.

- **Unsecured wireless networks:** Users are more likely to connect to unsecured wireless networks because of their nomadic nature and the ubiquity of wireless connectivity. This affords a very rich target for phishing-based attacks using a variety of attack vectors, such as browser-based, spurious *proxies* (rogue intermediaries that purportedly provide a legitimate function like a web proxy, but in fact are designed to steal information), SMS, and the like.

- **Typing errors:** Because of the limited real estate on the keyboard, users are prone to errors while typing URLs and therefore could be landing on spyware-infested websites that could launch phishing attacks.

- **Small-screen display:** The small screen size demands that the browser rendering of pages be optimized, and important information might be abbreviated or missing.

  - *Lack of security alerts and warnings:* On a small screen, detailed security alerts and warnings may never be rendered. Check your smartphone right now and try to verify the appearance of a website and its content.

  - *Lack of e-mail source headers:* E-mail clients often obfuscate the source headers of the e-mails for better rendering of the message. This kind of interface is ripe for phishing attacks because the headers are usually a dead giveaway for forged e-mails, and if this key indicator is missing, your users will be easily fooled.

  - *Lack of complete URLs:* See Figures 9-5 and 9-6 for the URL obfuscation that happens in portrait mode in an iPhone versus landscape mode, which happens to display the entire URL in this case. Even your most alert users are easy prey to a phishing attack when they browse in portrait mode because the URL isn't fully visible.

With this level of exposure to potential phishing attacks, it's critical that you have an antiphishing solution available. Antiphishing solutions for mobile devices can have a similar approach as the antivirus solution: All of it can be localized on the device itself, or you could take a hybrid approach by leveraging the hosted server in addition to installing a lightweight agent on the device.

Figure 9-5:
An iPhone in landscape mode with no URL obfuscation.

Figure 9-6:
An iPhone in portrait mode with URL obfuscation.

A variant of the hybrid approach is the cloud-based approach where the anti-phishing arsenal, e-mail, messaging and URL filtering, is entirely cloud-based. While this approach has a lot of appeal, without a smart agent running on the smartphone or device, an exclusive cloud-based approach falls short of the mark because of all the different interfaces it must maintain, which means that it has many different attack vectors.

For instance, even if the 3G interface is well-cleaned by the cloud approach, a local Wi-Fi or Bluetooth connection that is open can be used to compromise and delude your users to a phishing attack. Therefore, having a good *on-device agent* is key to providing that first line of defense against antiphishing attacks. (See Chapter 3 for more on cloud-based computing.)

Antispam

Antispam is the ability to identify and *stop* spam — typically in the form of e-mail — to the device, but note that for today's mobile devices, the spam vectors increasingly include SMS as well. From your users' perspective, the one distinct difference between e-mail–based spam and text-messaging–based spam is the latter sometimes costs your users (especially those in certain geographies; more on that in a moment) because some cellphone plans impose charges for messages received over a specified limit. Unlike e-mail spam, which is a major irritant (and a potential phishing mechanism), at least it doesn't cost the user money.

In response to this, carriers have been pretty active. In the United States, for instance, AT&T advertises a service called AT&T Smart Limits, which allows the user to block or allow text messages from certain users. Yes, it's an opt-in, paid service that users have to subscribe to. But according to research conducted by Ferris Research, in the United States users typically receive a couple of SMS spam messages per year, in Europe the frequency jumps to a couple per week, and in India it's a couple a day, while in China users are bombarded with 5 to 10 spam messages a day! So it's coming to a location near you. The geographical disparities in the SMS spam are a direct reflection of the SMS usage. In Asia and Europe where SMS usage is rampant, the spam frequency is also high, whereas in the United States, where SMS usage is relatively muted, the spam usage lags.

The other point is that there is a global uniformity component that needs to be factored in with any sort of solution you roll out. On the standards front, the GSMA (GSM Association), a consortium of nearly 800 members, has kick-started an initiative called *GSM spam reporting service* whereby users who receive spam can forward those messages to a standardized number. (It's currently proposed as #7726, which spells SPAM on the handset.) This is a neat way to build a database of blacklists for the spam operators and eventually use this information to build an in-network spam-blocking solution! Information about spammers will also be shared among participating members who will receive correlated reports with data on misuse and threat to their networks.

Antispam solutions — for e-mail or messaging — have more value if they're handled by the server rather than the client. This enables you to centralize the antispam solutions and apply remediation at the e-mail servers that you host — or apply it at your outsourced arm. For SMS-based spamming, the service is typically provided by the carrier, so you should actively work with your user's carrier, or educate your users about their carriers' services, to arrive at a solution that satisfies your needs.

A new variant of mobile spam is the use of applications on the mobile device to expose a new threat vector. For example, the Facebook app on your users' devices is one of the most popular applications in use. A clever spammer recently discovered a vulnerability to autoreplicate links so that unsuspecting users clicking any of the application spam links, shown in Figure 9-7, is enough to "share" (publicly post) the application on the user's Wall, and it spreads virally from there. Even though this isn't a mobile-specific spam vector, it's one that's growing in popularity using the social network applications for posting for spam and phishing attempts.

Figure 9-7:
Facebook
spam.

OH MY GOSH! I know I shouldn't talk about this here in facebook, but I thought I would message you from my new APPLE iPad that I just got for free. Don't tell anyone but there is a website sending out free iPad to anyone that signs up, Link to website (removed), that is where I got mine btw follow these steps .....

These kinds of social engineering–based spam are the hardest to mitigate and prevent, as these are predominantly tied to user behavior and tap into the psychology that the spammers become expert in exploiting.

You can fight Web 2.0–based spam more effectively by using the following:

- ✔ Constant vigilance
- ✔ Security posture adaptation
- ✔ Relentless education of your users

Using Backup and Restore Capabilities

Many smartphone OS vendors already offer some version of backup and restore. For instance, iPhone already comes with backup and restore capabilities whenever the device syncs with iTunes. But this is ultimately designed for end users, and the backup destination is anywhere the user chooses it to be. It also relies on diligent users who turn on this functionality in the first place. So this isn't something you can rely on. You need an enterprise-grade backup and restore capability that you can control.

A top-grade enterprise solution that RIM (Research In Motion) offers as part of their BlackBerry Enterprise Server automates backup and restore, as the BlackBerry Enterprise Server automatically syncs over the air with the BlackBerry devices and provides you with the ability to back up the BlackBerry Enterprise Server. (Typically, it's on secure premises.) In fact,

RIM is even extending this traditional enterprise server–based backup to the actual individual users so the users can take the matter into their own hands. Figure 9-8 shows one of the backup and restore management screens in the BlackBerry Protect user interface.

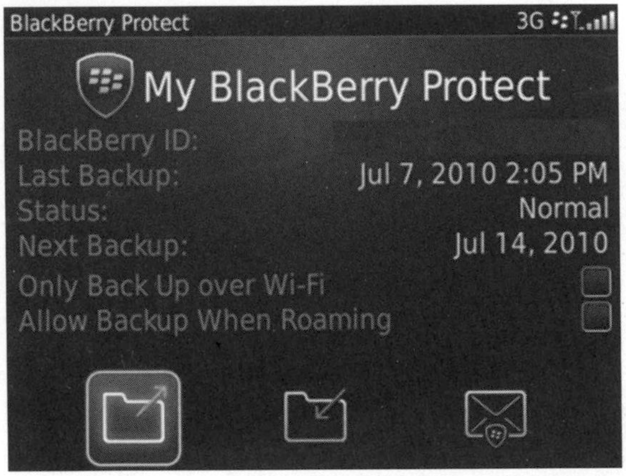

Figure 9-8: My BlackBerry Protect.

BlackBerry Protect	3G ⚡T.ull

🔰 My BlackBerry Protect

BlackBerry ID:	
Last Backup:	Jul 7, 2010 2:05 PM
Status:	Normal
Next Backup:	Jul 14, 2010
Only Back Up over Wi-Fi	☐
Allow Backup When Roaming	☐

The basic components of any backup and restore capability should

✔ Be able to do backups of smartphone data at a predefined frequency using over-the-air technology (as well as local backups when possible).

✔ Be able to do restore of smartphone data on demand using both over-the-air technology as well as a local connection.

There are a variety of ways you can provide this support, depicted in Figure 9-9 and explained as follows:

Figure 9-9: Backup and restore solutions.

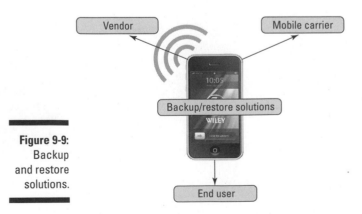

✔ **Vendor supported:** The BlackBerry fits nicely into this category, and very little mental exercise is required from you when you adopt this option. However, other smartphones and most devices don't support this option.

✔ **Provider supported:** Increasingly, carriers are starting to provide this as a service offering, and you may be able to capitalize on this by entering into agreements with operators and getting this provided as a managed service offering.

✔ **End-user supported:** This relies on end users regularly using the supported options to back up their smartphones. However, as noted earlier, the backup is typically local to their desktops and laptops only, so this solution in turn relies on your (hopefully) existing enterprise backup of their local machines to enterprise backup servers thereby backing up their smartphone backups. Wow! That sounds convoluted, and it is.

You should *not* adopt an end-user–supported option as your primary backup solution because it relies on end-user best practices, and while the workforce education is getting better all the time, relying on an informed workforce to guarantee backups is simply not recommended. Ultimately, you are responsible for protecting your company asset — the intellectual property — and need to exercise controls to do so. Therefore, you need to have backup solutions that can be scheduled, archived, and audited by you (and your stakeholders).

Adding Loss and Theft Protection

Your users are wedded to their mobile devices, perhaps more than they realize. A brief divorce from their beloved smartdevice is enough to cause heart palpitations and sweaty palms. These devices have become an extension of the owners themselves, so protecting them becomes a necessity — not a luxury!

The most fundamental defense against loss or theft of mobile devices is over-the-air (OTA) disabling. With enterprise-friendly devices like the BlackBerry, this is a breeze, but with most mobile devices, including the iPhone, iPod, iPad, Android-based devices, and others, this is a trickier proposition.

Thankfully, loss and theft protection is a rapidly evolving area, and all the leading device security vendors are rolling out various OTA device-disabling solutions to cater to this security need. Their antitheft solutions can be classified into these three broad categories:

✔ Encryption and authentication techniques

✔ Immobilizing techniques, including active data obfuscation following the loss of the smartphone

✔ Recovery techniques to locate the smartphone

The following sections cover each of these categories in greater detail.

Encryption and authentication techniques

As the name suggests, this technique obfuscates critical data on the device itself using encryption technologies. As you see in earlier chapters, extensible memory on the devices, including removable storage, makes the loss of the device quite dangerous. One mechanism that can mitigate this is encrypting the data on these memory cards so that in the event of a loss, the perpetrator can't access the memory card data using a card reader. Likewise, for onboard memory as well, using strong authentication techniques should be mandatory.

Your users will likely balk at the convoluted multilevel authentication techniques when you try to impose this on them and, worse, will always try to subvert this. You can never completely prevent this, so your best form of defense is education, education, education. In fact, you could use some provocative videos and scenarios where real users lose their devices and focus on the muted impact of someone who has followed the best practices versus a more damaging situation for a user who has grossly violated the encryption policies.

Immobilizing techniques

Here are the two most common immobilizing techniques:

✔ **Remote lockdown:** This technique involves an over-the-air kill message that is issued by the enterprise to the smartphone, which will essentially render the smartphone lifeless.

✔ **Remote wipe down:** This technique involves wiping out the critical smartphone data — contacts, local files, e-mails, SMS, and memory card.

Recovery techniques

These are the most common recovery techniques:

✔ **Smartphone locator:** Most of the modern day smartphones have a GPS chip built in. Using location software, the ability to track down the smartphone is becoming increasingly practical.

✔ **SIM snooping:** One of the first things that a stolen smartphone is subject to is swapping the SIM out. This provides an insertion point to use a technique called *SIM snooping,* which surreptitiously sends the newer SIM's telephone number to the original user, and this key piece of data can be used to locate the user with the carrier's assistance.

Carriers are getting into the act as well to provide protection against loss and theft. For instance, Verizon Wireless now offers to its customers the Mobile Recovery app shown in Figure 9-10.

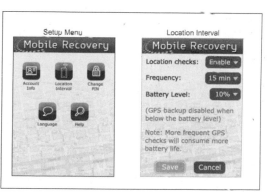

Figure 9-10:
Verizon
Wireless's
Mobile
Device
recovery
app.

If it's possible to strike up agreements with the key operators that service your locations, you may be able to provide carrier-managed recovery services.

Controlling and Monitoring Applications

Applications, or apps, are fast becoming the de facto user interface for mobile devices. Therefore, you need to be in sync with this trend and be able to provide adequate monitoring of these applications using various approaches (which we discuss in this section), identifying harmful applications in a timely manner and intervening when necessary.

Let's get real: Your users will download content (willfully or involuntarily) that is in violation of your enterprise policies. It's in your best interests — and your users' best interests, even though they may not embrace this notion right away — that you have good visibility into their application usage behavior and intervene where appropriate.

Be aware of any local regulatory matter that might forbid these intrusive policies, as in some regions they could be a violation of citizens' rights.

Methods to control and monitor applications

Now that you understand the importance of monitoring and controlling applications on your users' devices, you need to determine what type of solution you want to deploy.

There are two approaches to application control and monitoring:

- ✔ **Client-only:** In a client-only approach, you have a monitoring application running on every mobile device that you need to configure in the enterprise. While daunting, it provides you with an unparalleled degree of individual control, and you can set up policies that are unique to every user in the enterprise. More impressive is that you can take into account the real-time characteristics of the device — such as location, battery life, and other applications running — to make a much more customized strategy.

- ✔ **Server-based:** At the other end of the spectrum is a server-based approach that employs a centralized gateway to which all device traffic is backhauled and generic policies are applied. While user and device identification are still possible in this approach, and policies can be tailored to cater to the individual smartphone, the specific characteristics that an agent could supply in the previous approach are no longer available here. However, the economies of scale are evident, as you can have a centralized console for configuration, monitoring, and enforcement without having to worry about connecting to every individual device.

A more common hybrid approach is to tie in a lightweight agent with a server back end that can benefit from the agent providing the instrumentation and lightweight policy enforcement, with the server doing more complex application usage analysis and determination of policy changes that can then be relayed to the agent when appropriate.

Identifying harmful applications

You have to be on the lookout for seemingly harmless applications that your users download to solve a business issue. The application might seem innocent, but it could have an underlying security loophole that when exploited can cause all kinds of issues.

For example, an increasing number of new laws are mushrooming that ban automobile drivers from using their cellphones while driving. This has given rise to a number of text-to-voice applications that convert your text messages and e-mail into voice and play it back to you while you're driving. Seems like a very useful function. *Bad idea!* A number of these applications also use the "hybrid approach" whereby their app is actually a lightweight agent, and a bulk of the transcription happens in the cloud. So your users may actually be compromising valuable corporate data in the quest to be more productive while they're driving.

If you have an application-monitoring function in place, you can identify a harmful application by using the agent on the device, which would flag an unapproved application at install time. Alternatively, in a server-based environment, you can use tools to look for specific traffic patterns to identify corporate e-mail and texts that are going to unknown destinations and take appropriate action.

Enterprise Management of Mobile Devices

There is a lifecycle to which a mobile device in an enterprise needs to adhere to allow you a predictable process to get enterprise-ready mobile devices into and out of the workplace.

Enterprise management of mobile devices can be broken down into the following activities:

- ✔ Device deployment
- ✔ Device discovery
- ✔ Device provisioning
- ✔ Device monitoring
- ✔ Compliance enforcement

The following sections give you a complete overview of enterprise mobile devices lifecycle management. We delve into detail about each of these important phases of the mobile lifecycle.

Device deployment

Device deployment is relevant only if you intend to issue enterprise devices to your employees. If on the other hand (as is becoming the norm), the devices in your enterprise are predominantly owned by the employees themselves, this activity can be easily skipped.

An effective device deployment strategy involves providing a limited selection of devices from which your users can choose from (providing a large selection would entail a broader support burden that in most cases is unnecessary). Additionally, the strategy entails negotiating with carriers about bulk pricing (including pooling data, text, and voice across your user base) and shorter contract length to allow you sufficient flexibility to evaluate effectiveness and pricing periodically.

If, however, you're tasked with deploying devices in the enterprise, you need to take the following factors into consideration:

- ✔ **Device selection(s):** Opt for a limited set of devices to cater to a basic voice user as well as to the more advanced mobile device user. Device characteristics should reflect the enterprise policies we discuss earlier (device encryption, remote wipe capabilities, and so on).

- ✔ **Carrier selection(s):** Depending on the size of your enterprise and whether or not roaming is needed, your carrier selection may vary. Typically going with a larger carrier provides better worldwide coverage, but if you require local roaming only, some of the tier 2 and tier 3 carriers have very good promotions and service.

- ✔ **Pricing terms:** You need to negotiate bulk pricing that allows you to aggregate data, text, and voice across your users to give you better pricing based on consolidated usage and your users better flexibility by not having to carefully monitor their individual usage.

- ✔ **Contract lengths:** You should negotiate contract lengths down to the smallest possible terms to allow you periodic evaluation of effectiveness and pricing from the carrier.

- ✔ **Warranty terms:** Negotiate extended warranty terms for mobile device replacement as well as periodic upgrade terms because your users will expect reasonable upgrade cycles as technology and offerings advance.

Because the criteria used by similar-sized enterprises is about the same, you would be well on your way with device deployment by going ahead and copying, so to speak, from others. Look at competitors from your industry, and others as well with similar company sizes, to draw upon their device deployment choices.

Device discovery

Discovering mobile devices as they come into your network isn't a difficult activity to perform. Depending on how the device connects to the enterprise, there are various techniques to authenticate the user and the device. For instance, if a smartphone is using Wi-Fi to connect to the enterprise network, user authentication is a good first step to identify the employee. To further qualify the device, the ISAPI filters or Network Access Control (NAC) techniques can be used to zero in on the device type itself.

If the user is using the carrier's wireless network to connect to the enterprise, typically this is over a VPN connection like SSL, and the SSL appliance can be used to elicit the user and device information.

If the device isn't recognized, you can choose whether to block all access to the enterprise or limit the user access to noncritical systems. This is dependent on the strictness of your enterprise policy that you choose to enforce.

Device provisioning

Provisioning devices involves delivering configuration data and policy settings to the mobile devices. (Chapter 4 covers these policies in detail.) The leader in this space is clearly Research In Motion, with its BlackBerry Enterprise Server that, with a few mouse movements, can remotely configure these devices. Other device and OS vendors are catching up to this crucial enterprise requirement. For instance, Apple's configuration and security settings can be deployed wirelessly through a user self-enrollment portal. While not as transparent to your user community as the BlackBerry solution, it does provide similar levels of granular configuration and policy deployment to iPhones and iPads.

Note that provisioning may be a one-time activity, but keeping up with OS upgrades, security patches, and so on is always going to be needed on an ongoing basis. The good news is that with employee-owned devices, the upgrade itself is the onus of the employees, and the device and OS manufacturers have done a pretty good job at making this as painless as possible. But you need to be prepared for users still fighting these upgrades. You must be able to modify your policies to allow or disallow access to the enterprise, taking into consideration the revision level of the operating system and applications so that your risk level isn't increased due to nonconforming devices coming into the network.

Device monitoring

Device monitoring is a constant activity that's needed to ensure that the mobile device is in compliance with your enterprise policies at all times. Note that this is different from application control and monitoring, which have more to do with user behavior and being able to control what the end user is trying to accomplish. For more information on application control and monitoring, see the section "Controlling and Monitoring Applications," earlier in this chapter.

The primary objective of device monitoring is to validate device compliance to enterprise policies at all times.

Typically, an enterprise agent is installed in the provisioning process (described in the preceding section), and it's your eyes and ears for every enterprise mobile device. This agent has some unique characteristics — it needs to consume very little CPU and battery power, and it needs to be as unobtrusive as possible until a violation is detected.

The CPU and power budget consumed by this agent is critical because your users are going to revolt if this agent causes their smartphones to become unresponsive or drain significantly. Therefore, any vendor that you choose needs to demonstrate a lean and mean profile that can perform the tasks unobtrusively.

And *unobtrusively* means that when the device discovers a known enterprise wireless access point and connects to it, the device monitoring agent monitors this connection to ensure that all communication thereafter stays encrypted.

Compliance enforcement

The next step in the flow, once your device-monitoring agent detects a violation of your enterprise policy, is enforcement, which comes in these three flavors:

Your primary goal is protecting the enterprise from any security breach caused by the errant application. However, it is also important to ensure that your users have the most painless experience possible as you try to remediate the situation. To this end, the more transparent this reparation is to the end users, the more satisfied your users will be, and ultimately, you will breathe a whole lot easier as well.

- ✔ Automated correction that's transparent to the user
- ✔ Automated correction (through redirection to a self-service portal) that requires user interaction
- ✔ Manual remediation that requires your intervention (highly undesirable — see the following warning)

The last option of manual remediation should be your "have no other recourse, therefore I am going to resort to this" option. This is a highly intrusive, labor-intensive, and user-unfriendly option. But in certain situations, this may be the only option available to you, so we cover this as well in the subsequent sections.

Automated correction

As the name suggests, automated correction is the most painless of approaches, as it doesn't require any intervention by you or the user, and the device automatically self-corrects. This intelligence is embedded in the monitoring application itself.

For instance, consider the previous example for device monitoring where the smartphone connects to an enterprise wireless access point and the monitoring agent's job is to verify that all traffic is encrypted. When a device is found to be in violation, it could be easily self-corrected to enable encryption on the smartphone (IPSec or equivalent) without affecting the end user and still ensure that your enterprise policies are met at the same time!

Semi-automated correction

The semi-automated remediation capability requires the user to be involved. Typical violations are applications that are downloaded that violate enterprise policies. An example of this could be an application that provides cloud storage as an extension to local device storage. Clearly, for data that is stored locally, you would have policies such as local encryption, but once this data extends to the cloud, there's no way you can enforce such policies.

Your only resort at that point is to disable these classes of applications. However, since you can't willy-nilly delete applications on your employees' devices, your only resort is to redirect them to a remediation portal where they're presented with the facts. Perhaps, a notice such as this would work:

> "Your Foo app is in violation of enterprise storage policies. You will not be allowed access to the enterprise network until you disable or delete the Foo app from your device. Have a nice day."

Be succinct yet comprehensive so the user is faced with the choice of either deleting the app or choosing to retain the application but not connecting to the enterprise any more.

In either case, there is active user involvement, and giving users a choice offloads the burden from you and your staff having to deal with these errant users!

Manual remediation

Manual remediation is the most intrusive as it involves you — the IT department — having to play a part in enforcing the enterprise policy. Typically, this happens if automated correction isn't possible or there have been recurring violations that need active intervention on your part.

An example where automated or semi-automated correction isn't possible is when a new vendor's device is introduced to the network. As discussed earlier, you have a choice of blocking all access or limiting it to noncritical systems. But a third choice is to meet with the user to learn about the device and its capabilities — and maybe even add that to your catalog of supported devices and adapt your policies based on this device's unique capabilities. This keeps you ahead of the game, and more importantly, you are viewed as a trusted and flexible partner who is constantly working with your user community to make them more productive!

There may be employees who are constantly looking for loopholes to violate enterprise policies. They experiment with jailbreaking, downloading rogue applications, compromising security on the network — the list goes on. Your options with these repeat offenders include temporarily suspending access to the enterprise network and, if that doesn't get them to wake up and smell the coffee, escalating to your management and their management, when that's the only recourse. It happens; so be prepared for it.

Chapter 10

Hacker Protection and Enforceable Encryption

In This Chapter

▶ Knowing the components of on-device security components

▶ Protecting devices with device-based firewall software

▶ Fighting viruses with device-based antivirus applications

▶ Warding off spam with device-based antispam

▶ Guarding against device intrusion

▶ Defending your communications with device-based encryption

This chapter covers each of the on-device Anti-X protection components, first covered in Chapter 9, in detail. The two important concepts that are the focus of this chapter are hacker protection and enforceable encryption. We discuss various on-device protection capabilities that can be effectively used to defend against hackers. Additionally, you find out the importance of enforceable encryption and discover the tools and techniques available to enable this critical piece of your mobile security arsenal.

Getting to Know the On-Device Security Components

On-device security components are firewall, antivirus, antispam, intrusion prevention, and enforceable encryption. We discuss the benefits of using each of these on-device security components in detail throughout this chapter. When you roll out on-device security components, you must prioritize: Which components do you implement first? (In this book, we call those the "do-these-first" components so they're easy to spot.) Eventually you want to arrive at a complete security strategy that includes *all* the components in your security arsenal, but let's get real: First things first. You need a horse in front of the cart to get things rolling.

In this chapter, we explore the following on-device security components at length (see Figure 10-1):

- **Device-based firewall:** A native firewall running on the smartphone as a barrier against uninvited intruders.

- **Device-based antivirus:** Antivirus technology installed on the smartphone.

- **Device-based antispam:** Built-in filtering capability that protects your users from the unwanted barrage of incoming bulk e-mails.

- **Device-based intrusion prevention (including SMS):** A sophisticated and application-aware firewall designed specifically to detect attempted hacks and protect the smartphone against intruders.

- **Device-based enforceable encryption:** A feature that protects data communication on the smartphone by obfuscating the data resident on the device and encrypting data in transit as it originates from the device.

The threat vectors are many, and the forms of attack are varied. Any solution you adopt must be versatile enough to counter this multi-pronged frontal assault that users and their devices face day and night.

Your users' biggest nightmare (at least the one that haunts their smartphones) is battery life — or lack of it. Turning on all the on-device security solutions at once may be tempting, but it hits battery life big-time. Therefore, you have to pick and choose a combination of on-device and provider-based models for a more practical deployment solution.

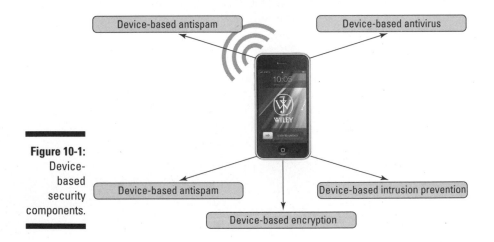

Figure 10-1: Device-based security components.

Keeping Devices Safe with On-device Firewalls

A device-based firewall is a form of protection that is physically resident on the device, as opposed to protection based in the cloud or hosted protection. A device-based firewall's express purpose is to detect and thwart relatively straightforward brute-force attacks.

A firewall will typically thwart unauthorized external connections that attempt to communicate with the device. The firewall can even be configured to monitor (and block as necessary) internal applications on the device that attempt to communicate with the outside world.

The adoption of firewalls for smartphones is becoming more mainstream mainly because these phones look and feel increasingly like laptops and desktops, and everybody's familiar with client-based protection that includes a firewall for laptop and desktop computers. So the argument in favor of extending similar types of protection to cover smartphones does pass the smell test — yes, it's necessary — but it's not sufficient. That's partly because (also increasingly) these smartphones are becoming users' primary means of connecting to the Internet. No wonder the exposure level continues to skyrocket for smartphone users as they spend ever-increasing amounts of time online, as shown in Figure 10-2.

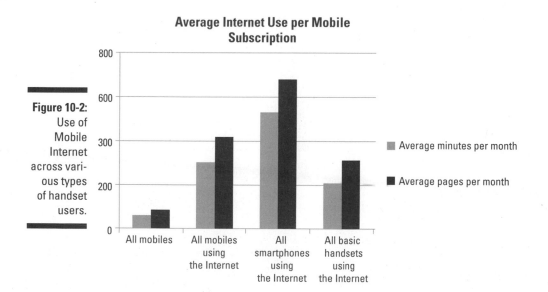

Figure 10-2: Use of Mobile Internet across various types of handset users.

Source: GSMA Mobile Media Metrics

Later in this chapter, we cover device-based *intrusion prevention* (protection that watches out for more advanced and sophisticated threats). You can use the firewall *and* intrusion prevention in tandem to get the most comprehensive protection for your users. Be warned, however, that this approach increases battery drain: The more intensive the protection that intrusion prevention provides, the more processing power it requires. The recommendation here is to use on-device intrusion prevention solely as a backup option if you see that your users are getting attacked mercilessly and you need the heavy hammer to protect them.

The device-based firewall, on the other hand, provides basic protection against common attacks and therefore demands less power. So the question is: How much protection do your users need, and how much battery drain will they tolerate?

Although smartphones are similar to laptops and desktops in many ways, the differences are striking, and the security posture you adopt must take those differences into account. The following sections look at some of those crucial differences: the footprint of the device, battery usage, and adapting to changing usage patterns. Keep them in mind when you adopt a device-based firewall for your smartphone users.

Small footprint

It's critical to choose a firewall with a small footprint — that is, a modest use of power, storage and memory — on the smartphone. Even as storage gets cheaper, your users' smartphones become voracious animals gobbling mega- and gigabytes by the minute, and much of the available space is already taken. Therefore any business-critical application that you want to run — the firewall clearly fits that bill — must be small enough not to interfere with your users' lives. It should go about its tasks as unobtrusively as possible.

The striking similarities between the smartphone and its laptop/desktop predecessors have not gone unnoticed by the traditional firewall vendors — TrendMicro, Symantec, and the like — so all of them offer versions of their products for the smartphone. When you choose a firewall, scrutinize each product to see where the vendor has drawn upon heritage solutions — where size was not the biggest concern — to deliver a smartphone version of the same product. The challenge was to get the footprint down; how well does each product succeed?

Determining the memory footprint of your firewall application for your mobile device is not a straightforward process. If you use a system monitoring application like the System Activity Monitor for the iPhone or System Monitor for the Android, you can see the current memory usage. Subsequently install the firewall application and run the same system monitoring application again and note the new memory usage. The difference between the two is the memory footprint of your firewall application.

Contrast this with some of the "mobile only" firewall vendors, such as Lookout and SMobile (now part of Juniper), who have had the luxury of designing their smartphone solutions from the ground up. Their products are optimized for Internet-connected use but may lack some advanced firewall features that the traditional players have had for years.

All in all, it's a potpourri of vendors and solutions out there, and you have to balance the "small footprint" requirement against the "best features" and "most bang for the buck" criteria when you evaluate their offerings.

You first want to ensure that your enterprise policy is being adequately met in terms of the features that the firewall vendor provides. The footprint should be your next immediate consideration and not the other way around.

Some vendors may have optimized their solutions for particular operating systems or specific devices. Make sure the solution you choose fits what your people are using. Identify the top two to three smartphones in your enterprise and run the "footprint" test against all of them. You don't want to be lulled into a false sense of security by a device-specific solution, only to be rudely surprised later. Be sure to do your homework thoroughly.

Efficient battery usage

Battery usage applies to every single application that runs on the smartphone, but it has a special significance to the on-device firewall: It must run as efficiently as possible and consume the least possible battery power.

For users, convenience is a priority; they'll turn off applications that they think are draining the battery too much. You don't want the firewall to be one of the offending programs that gets identified as a power-hog. You can run a device-based monitoring agent to alert you if the firewall is deactivated, but frankly, that should not be a frequent occurrence. Your firewall should be efficient enough to keep it off the list of things-to-turn-off; otherwise you'll age rapidly in your day job.

Battery life — or the lack of it — is a constant nightmare for your user community. Therefore, it's in everybody's best interests that you choose an efficient firewall to put on your smartphone. It's akin to checking the footprint when you're evaluating operating systems and devices: Does the product use only the storage and memory that it needs — and no more? The same holds true for battery life: the firewall should do its job with minimal power drain.

You need to learn from the experience of others when it comes to battery usage of firewall vendors because there is no easy way to glean this beforehand by yourself. Actively scour the Internet and trade magazines to look for actual user experiences and reviews before making a decision.

Dynamic adaptation to changing usage

Keeping your security response flexible is uniquely important to the mobile environment. That's partly because most current smartphones make multitasking available. Your users could be videochatting with one application while simultaneously texting, make a voice call, turn on location-based services to find the nearest gas station, and download corporate e-mail — all at the same time. Any firewall that claims to protect the device has be able to watch constantly for the specific applications, interfaces, and protocols the user is using *at any given moment* and provide complete protection against attack for all of these.

The heavy-hammer approach is tempting: You could turn on protection for all interfaces, applications, and protocols at all times but then the firewall falls afoul of the "efficient battery usage" tenet. You don't want the firewall to suck the life out of the battery while trying to protect everything constantly, regardless of what's actually in use. Clearly, an effective firewall has to be more intelligent, adapting constantly to the usage pattern and turning protection on and off as necessary.

In terms of the types of interfaces a firewall needs to protect, it comes down to what types of wireless connectivity your mobile devices provide. Typically, a mobile device has at least a wireless LAN or Wi-Fi interface that allows the user to connect to the wireless network. In addition, for smartphones, the device has an interface (like a 3G or 4G interface) to connect to the service provider's network. Most firewalls provide protection against these two primary interfaces, and you should make sure beforehand that they do. In addition, you need to consider other interfaces your mobile device may have, including a Bluetooth interface. A Bluetooth interface is particularly vulnerable because not many firewall vendors protect this interface, and as you are well aware, it is one of the most widely used interfaces for accessories like a headset, Bluetooth stereo devices, and so on. See the nearby sidebar for an in-depth look at the Bluetooth security issues.

Understanding the vulnerability of Bluetooth

One particular area of vulnerability for the mobile device is the Bluetooth interface. Traditionally, Bluetooth has been an adjunct communication interface used for connecting to wireless headsets and keyboards, to the smartphone integration system in cars, and to other such accessories. More recently, Bluetooth has become a conduit for Internet connectivity using a technique called *tethering* that allows the mobile device to function like a "modem" through which your desktop or laptop can connect to the Internet. All in all, the hitherto-unsung Bluetooth interface is becoming more prominent for your users. Keep in mind, however, that most device-based firewalls typically cover all IP interfaces — WLAN, GPRS/EDGE, 3G, LTE, and the like — and they may not provide specific coverage for the

Bluetooth interface. And if your users fire up Bluetooth, assuming coverage while blissfully unaware of the vulnerability, they may be lulled into a false sense of security. ("Hey, the fire-wall on my smartphone has me covered, right?" Well, no.)

The first recommendation you may want to make to your users is, "Turn off Bluetooth." Reality check: That's not practical, and even less likely to be followed. But even the National Institute of Standards and Technology (NIST) took up the refrain when it issued its "Guidelines on Cell Phone and PDA Security," recommending that the user actually "curb wireless interfaces." The idea was for users to turn off any interface they weren't using until those interfaces were actually needed. Here's another reality check:

The majority of your users most likely favor the convenience of "always-on" interfaces. They'd rather not go through the trouble of turning those things on and off "as needed," and that's unlikely to change. Clearly you need a more pragmatic solution.

Optimistically, it's only a matter of time until Bluetooth interface security shows up among the features that on-device firewalls offer. Until that happens, you can include a Bluetooth-specific firewall, *in addition to* the on-device firewall, as part of your recommendation. For instance, Fruit Mobile offers a firewall for Android devices that protects specifically against Bluetooth attacks, as shown here in the figure.

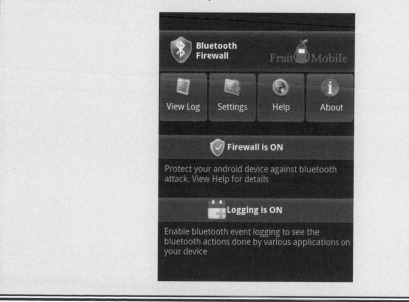

Protecting Against Viruses

With the widespread use of applications that download attachments to the mobile device — including the most widely used app of them all, e-mail — the need for virus-based protection is becoming critical. Keep in mind, however, that other mobile-specific *attack surfaces* (exposed areas that are vulnerable to attack by hackers) allow for other ways to infect the mobile. For instance one of the earliest mobile viruses was the Commwarrior that would propagate itself by using MMS message attachments and Bluetooth. Once the virus infected the device, it would start searching for nearby Bluetooth phones to infect.

These attacks use features found specifically on mobile phones — MMS, Bluetooth, and the contacts database — to compromise the device and propagate the attack. Figure 10-3 shows three fundamental ways such an attack can be launched to propagate a virus to a smartphone.

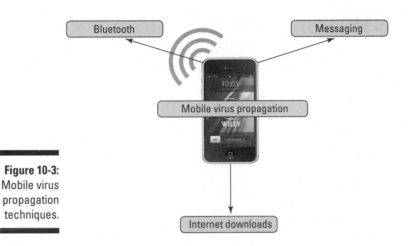

Figure 10-3:
Mobile virus propagation techniques.

Here's how each mobile virus propagation technique works:

 ✔ **Bluetooth:** This technology has come of age with the widespread use of hands-free devices as well as close-range device-to-device communication. The standard operating configuration for most users is to place the device in *discoverable mode* (so it can be seen by other Bluetooth-enabled devices nearby) or *connected mode* (which allows the device to be discovered *and connected to*). Viruses can be delivered to the device in either of these modes.

Note that this potential risk can be overcome by completely turning Bluetooth off so you eliminate the Bluetooth attack surface. However your users are not likely to do so because it's not user-friendly. They're much likelier to keep their phones both "discoverable" and "connected," which makes them sitting ducks for virus attacks.

 ✔ **Messaging:** Malware attachments can be appended to messaging services such as e-mail, MMS, or Instant Messaging. Typically the default configuration does not allow these attachments to unpack and run automatically; the user has to accept the attachment *and* open it and become infected. But you probably know the dazed look in your users' faces when they see those deadpan warnings; expect some of them to ignore the warning and fall victim.

> ✔ **Downloads:** This is probably the most widely used way to disguise and deliver malware. All the device needs is an Internet connection; the incoming malware-infected file can show up disguised as (say) a game, security patch, software upgrade, utility, shareware program, video, picture, you name it. Even worse, an infected server from a reputable vendor can cause even the most cautious users to become unsuspecting victims to file-based viruses.

Antivirus technology has been available for decades, and many of your users would never consider operating a computer without some antivirus solution running on it. Antivirus protection is necessary — okay, they get it. But they only think of it as needed for their desktop computers.

A lot of us don't seem to notice that most of our mobile devices, which are all derivatives of computers in one way or another, have no antivirus protection whatsoever. What's even more surprising is that your users attribute more importance and personal attachment to the smartphone than to the computer while still failing to protect that phone.

Falling victim on social media

A variant of the traditional Internet download is the social engineering–driven malware download. For instance, if your users see that their online friends are downloading a particular Facebook application, they feel compelled to do the same. They figure they can trust their friends to download "safe" applications ("Well, duh! They'd never do anything, like, *unsafe* or something, right?"), and they don't want to miss out on all the fun and action their friends must be having. Using this kind of ingenious social engineering, malware authors can easily penetrate a social group and then watch the "fun" as users self-inflict injury, confusion, damage, and pain. One nasty and clever worm called "Koobface" (a play on the name "Facebook") was written specifically to target Facebook using components like these:

- ✔ Koobface downloader
- ✔ Social network propagation components
- ✔ Web server component
- ✔ Ads pusher and rogue antivirus installer
- ✔ CAPTCHA breaker
- ✔ Data stealer
- ✔ Web search hijackers
- ✔ Rogue DNS changer

Looks like a list of features for a sophisticated product, doesn't it? And that's exactly what Koobface is, except this "product" is nefarious, written with the express purpose of infecting accounts, propagating, and stealing Facebook users' information and identities.

You have to take a stand on this issue and ensure that you're providing adequate mobile antivirus coverage to your users on their both their desktop *and* mobile devices. Fortunately, the range of mobile antivirus solutions is ever-increasing. As with traditional antivirus solutions, you should be looking not only at the upfront costs, per-seat license renewals, and automatic signature updates, but also at mobile-specific features such as battery-life recognition, memory requirements, and the broadest possible coverage of mobile operating systems.

The following sections cover the types of antivirus solutions that are at your disposal. We also clarify the difference between the firewall (discussed earlier) and the antivirus solution and why you need both in your toolbox. As noted in Chapter 9, one common on-device antivirus solution uses the client-server model: An on-device agent program leaves the heavier processing to the server in the cloud. In the following sections, we revert to a more traditional on-device antivirus solution where all the processing happens on the device itself.

Firewalls and virus-based attacks

The previous section, "Keeping Devices Safe with On-device Firewalls," considers the device-based firewall as protection against attacks directed at the device itself. These attacks take various forms; two typical ones are

- **Port scanning:** The attacker looks for exposed ports that could be used to connect and compromise the device.
- **Brute-force ping floods:** The attacker barrages the device with pings to overwhelm its capabilities.

Most of these attacks try to exploit poor security postures of particular devices and applications. The perpetrator usually doesn't pay much attention to the smartphone's operating system.

Virus-based attacks, on the other hand, are more general. They're essentially file-based; they ride in on a file that must be downloaded (either overtly or covertly) before the attack can be launched. That's where an obvious operating-system concern enters the equation and becomes extremely relevant: Any device with the targeted operating system, mobile or not, can be vulnerable. Some of these viruses also target browsers, taking into account the browser and operating-system vulnerabilities.

A bevy of device-based antivirus solutions has popped up, and more come to market by the hour. The traditional desktop and notebook vendors (Symantec, Trend-Micro, Kaspersky, and the like) have morphed their offerings to support the newer smartphones. On the other side are the new kids on the block — vendors of smartphone security such as Lookout, F-Secure, and such — who provide highly customized smartphone antivirus products.

Which vendor(s) you choose depends on what's most important for you to protect as you explore this new dimension of your network — the smartphone. For example . . .

- ✔ If your predominant disposition is toward a common look and feel and consistency across all endpoints (desktops, laptops and smartphones), then you should look to one of the traditional vendors. As mentioned earlier (see the "Small footprint" section), familiar antivirus products that have a large footprint in the desktop environment have extended themselves by getting small enough to fit into a smartphone.

- ✔ If your primary goal is to provide a customized and tailored smartphone-centric antivirus solution, then you would do your due diligence (*and* do yourself a favor) by checking out the new-age smartphone antivirus vendors and choosing a mobile-centric product to fit your environment.

Virtual device antivirus solutions

For the sake of completeness, here's a look at virtual devices as antivirus solutions (as mentioned in Chapter 9). A "virtual" antivirus solution doesn't run on the smartphone itself; instead, the main program runs elsewhere on the Internet, making its features available through a small software agent running on the smartphone.

Here's how it works: The user downloads an antivirus agent to the device, and the bulk of the intensive antivirus processing takes place on a remote server (either locally hosted by you or by a hosted cloud service). The client collects information about the mobile device it resides on, and delivers a certificate of authority. In this model (shown in Figure 10-4), you maintain a clone of the actual phone in the enterprise as a virtual machine; the agent informs you of any changes to the end device — such as new applications installed, SMSs received, and so on — and then syncs with the virtual phone in the enterprise.

Any virus-based attack that is launched is actually targeted at the virtual smartphone, and the heavy burden of detection and cleansing is all performed in the virtual server. The smartphone itself, for all intents and purposes, is oblivious to the attack.

Note that you do need significant restrictions on the smartphone itself, such as not opening up any other interfaces (like the Bluetooth interface discussed earlier) because the only conduit from the smartphone needs to lead to the virtual smartphone and nowhere else. Opening up other interfaces on the smartphone could lead to directed attacks on the device, which renders such a "virtual" solution useless.

This is not real-time protection of the actual device, true, but it's reasonably close. And it has the advantage of not dragging down the smartphone's performance or draining battery at one gulp. In addition, because the capability is hosted on a server, you have a lot more processing power available for antivirus checking, as shown in Figure 10-4.

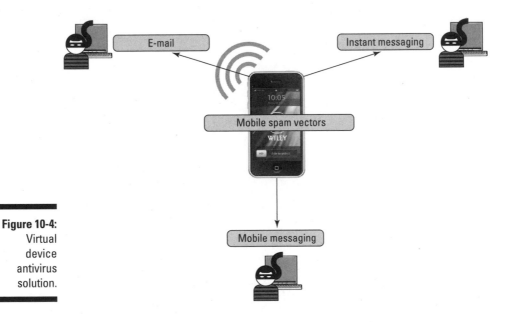

Figure 10-4:
Virtual device antivirus solution.

Reducing Spam

The threat of spam is as prevalent for mobile devices as it is for fixed devices such as laptops and desktops. This age-old form of malware continues to plague consumers and enterprises (you and me) alike. There are three primary places spam can come from when its target is a smartphone.

Here is a description of each of the vectors of mobile spam:

- ✔ **E-mail:** The most common way to launch spam is via e-mail. Although this kind of attack is not limited to smartphones by any stretch, the increased adoption of smartphones — and the gradual shift toward using mobile devices for primary e-mail connectivity — makes spam-clogged Inboxes a real (and likelier) concern.

- ✔ **Instant Messaging:** Attacks that use Instant Messaging — already a threat to traditional computer networks — are now more common on smartphones. Large communication providers and OS vendors offer not only the familiar form of Instant Messaging but also access to Twitter,

Facebook, and other social media, which are also instant communication channels. As discussed in Chapter 9, social media spamming is one of the most dangerous threats to your users because social media resonate with them more closely than do other forms of communication, and their defenses against this type of spam are practically nonexistent.

The most important way to counter social media spam is the same way you counter other threats — with a three-pronged defense:

- Be vigilant

- Adopt a security-oriented posture

- Relentlessly educate your users

✔ **SMS and MMS:** The mobile environment has its own unique form of spam based on mobile messaging, in particular SMS and MMS. As any employee who's used a mobile phone abroad can attest, hordes of spam SMS messages can hound the user to a disturbing degree. What's even more jarring is that in quite a few places, incoming SMSs are charged to the receiving party, so now the user not only gets an Inbox full of uninvited mobile spam but also has to pay for it.

While the threat vectors (ways to get spam to your device) can vary widely, the intent of the perpetrator(s) remains the same:

- Entice users to part with their money by making grandiose marketing claims.

- Phish for users' data (or simply trash their devices) by getting them to open the message and click following links that load malware.

Service provider assistance

As mentioned in Chapter 9, the bulk of antispam solutions are provided by the hosting entity (e-mail, service provider, content provider, and so on), and the reason is simple: Identifying and stopping spam *before it gets to the device* is the most efficient way to counter this threat. In addition, you can identify spammers by utilizing a large aggregation of information — whether in the form of e-mail or messaging — at point-of-service provisioning. Then you can constantly update and hone this database to make the blacklist more effective. Real-time protection can only be achieved by service providers who have the data, computational resources, and means to adopt this approach.

Choosing an antispam solution

A plethora of device-based solutions abound that claim to stop spam in its tracks. While this claim may be more marketing-speak, it's clear that to protect your users effectively against all the various spam threats, your solution needs a device-based component for spam protection. Although such a component should not be the only antispam solution used in your enterprise, it's important to include in your tool chest.

As we've seen with on-device firewall and antivirus solutions, the choice of a particular antispam vendor is based on a number of factors, but be sure to keep these particular points in mind:

- Which smartphones the product supports
- Whether the product works with existing antispam solutions in your enterprise
- Whether you can get enterprise-wide policy support for the solution
- Whether a cloud-based antispam solution is available and works with the on-device component

Human nature being what it is, be prepared for a small portion of your users to fall prey constantly to spammers. If antispam protection is one of your responsibilities, have a well-thought-out remedial action ready to take and a well-articulated recovery procedure in place. It's time well spent.

Global operator initiative to combat spam

With the widespread advent of mobile spam, the GSMA (GSM association) — the largest mobile consortium of its kind with nearly 800 members — has taken matters into its own hands. It has also kick-started an initiative called *GSM spam reporting service* whereby users who receive spam can forward those messages to a standardized number (it's currently proposed as #7726, which spells SPAM on the handset). This is a great way to build a database of blacklists for the spam operators and eventually build an in-network spam-blocking solution. Information about spammers will also be shared among participating members, who will receive correlated reports with data on misuse and threat to their networks.

A successful trial of this service concluded in December 2010, and the service is now available to operators worldwide to join. Customers can report spam they encounter and help build a robust and growing database that can be used by all operators worldwide to stop the advent of this nuisance.

Preventing Intrusion

This section builds on the previous section, "Keeping Devices Safe with On-device Firewalls," and delves into the dark and murky world of even more intrusive attacks and the armory you have to protect your unsuspecting users: on-device intrusion prevention. But before we delve in, keep in mind that intrusion prevention is computation-intensive. It takes processing power and (you guessed it) battery power.

The more intensive the protection that intrusion prevention software provides for your users, the greater the cost in processing power, which translates into the single biggest fear that mobile users have: battery drain.

In essence, you have to counter intrusive, high-tech attacks against smartphones with low-power-but-effective weapons.

There are three primary mechanisms that provide vehicles with which mobile devices get infected. These are

- ✔ Installing malicious apps (or seemingly innocuous apps that are actually malware)
- ✔ Connecting to "untrusted" networks, typically over wireless LAN
- ✔ Getting infected with spyware using SMS

In the following sections, we examine each of these mechanisms in turn, explaining how an infection happens and how to prevent against it, or at the very least detect and remediate.

Installing malicious apps

From a hacker's point of view, a smartphone presents a target far more fertile and wide-reaching than what a traditional desktop environment provides. Why? One primary reason — the app store. Thanks to Apple's revolutionary, easy way to download free and cheap (and some not-so-cheap) apps to smartphones, apps proliferate for every type of smartphone out there. (See Figure 10-5.) With hundreds of thousands of applications available to your employees with a touch of a button (and sometimes a voice command, so not even touch is needed), it's no wonder they constantly experiment with new applications.

"New" does not necessarily mean "good." Some applications are written with the same rigor as traditional desktop applications, but most are written with the express intent of getting a quick return on investment. Bottom line: Their creators tend to circumvent good programming techniques, which makes them vulnerable to attack.

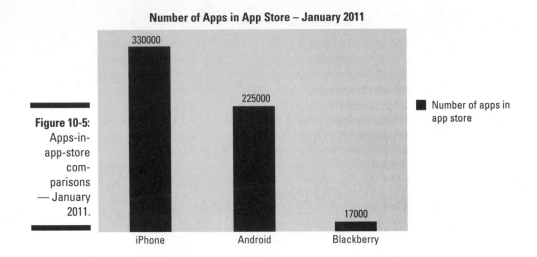

Number of Apps in App Store – January 2011

But wait, there's more; these applications, vulnerabilities and all, could also have widespread access to your employees' data stored on the smartphones, (contacts, messages, location, photos, and so on). An attacker could compromise those as well — and compound the trouble.

The familiar patterns of attack reassert themselves. One application that was sold as a simple wallpaper program also sent stored telephone numbers to a Chinese server; malicious Trojan Horses have turned up in gaming applications. One hack forced phones to make long-distance international calls, so the phone owner was then heavily charged for the "privilege" of being hacked.

When you're responsible for a device in the enterprise, you have to include all the associated applications, data, and security posture that go along with the smartphone or device. The relevance of the application explosion to these concerns is that it's also an explosion in potential smartphone vulnerability, and by extension, any associated data that a smartphone's apps may have access to is also vulnerable.

So what can you do to prevent against this? The odds are stacked against you — users will experiment with new apps, and malware developers will flock to generate nefarious apps that threaten to expose data, become part of a botnet, or wreak havoc on other devices. Your best tools for prevention are education and communication. Constantly and consistently communicate about the need for users to employ good judgment when downloading apps. Users should follow these practices:

✔ Avoid downloading apps created by unknown individuals.

✔ Check the rating and feedback provided by other users on those apps before downloading any apps.

✔ If the apparent value that an app touts sounds too good to be true, then it probably is. Avoid these apps at all costs.

After you pick your on-device firewall vendor, check with the vendor to see if it provides outbound monitoring solutions as well, which could point to anomalous behavior on the mobile device that could, in turn, provide insight into errant applications.

Connecting to unknown networks

As we have seen repeatedly, the mobile nature of the smartphone further exposes it to attacks that a fixed device may not be subject to. Here's why:

✔ A mobile device is always on the go.

✔ Smartphones support a plethora of interfaces.

Bottom line: The likelihood is very high that any given smartphone is connected to one or more wireless networks almost all the time and could be anywhere.

This constantly nomadic existence and propensity for tethering means much greater exposure to unknown networks. Therefore, intrusions are far more likely on these devices than on a fixed desktop.

Typically, these intrusions are in the form of an infected machine in the network. Note that nomadic users connecting to ad hoc networks — sometimes with no encryption on the network — present a very tantalizing target for hackers. Hackers could be on that same network with an infected machine or could have infected a device that they're controlling remotely from their console with the express purpose of attacking unsuspecting users as they attach to these networks.

So what can you do? Education, education, education. Your users need to follow these guidelines:

✔ Check for the security posture of a wireless network you are connecting to. Is it encrypted (WEP, WPA, WPA2, and so on)? If not, understand that there are risks associated with connecting to this network.

✔ Use the company-provided VPN client to ensure that all traffic is encrypted.

✔ Run the antivirus/firewall client after you log off from the network to look for any potential breaches that may have occurred.

Your panacea is to periodically scan your enterprise-issued mobile devices using the antivirus, firewall, and newer forms of threat-detection solutions as they emerge, assuming your users will be constantly coming on and off public networks. For non-enterprise-issued devices, there is little you can do in terms of exercising control at the endpoint, so your posture should be more defensive, looking for threats emanating from the mobile that could attack the enterprise and using your network-based security solutions to prevent against this. Chapter 9 details enterprise security policies that you can adopt to mitigate this.

Getting infected with spyware via SMS

One of the most popular applications on the mobile device is SMS — and this popularity has not been lost on the hackers! As mentioned earlier — and it bears repeating — spyware is a real threat. One particular variety manipulates SMS messages and exposes them so they can be read by others in the near vicinity.

Given the ubiquity of SMS, it provides a compelling attack vector for hackers to exploit, and an innocuous-looking SMS can result in an inadvertent installation of spyware by the unsuspecting user. This spyware then scans through the contact database of the user and starts spamming the user's contacts via any means possible, including SMS, e-mail, or even a call to the contact.

So what can you do? Again the same refrain — education, education, education. You users need to adopt these practices:

✔ Be wary of any unsolicited SMS and don't click any embedded links in the SMS.

✔ A rapid drain in battery usage should serve as a warning that spurious applications may be at play. You need to get the device inspected by a technical expert. If it's an enterprise-issued device, contact IT; if it's a personal device, contact the service provider or other reputable third-party.

You need to be vigilant in keeping up with the variety of SMS-related threats prevalent in the mobile device ecosystem. At a minimum, work with the service provider for the enterprise-issued handsets to negotiate a service contract that includes a device replacement policy as well as a service for cleansing infected devices of spyware. The provider may also offer an SMS-related security service that you should consider seriously if there is an uptick in user complaints.

Using Enforceable Encryption

One way to counter spyware is with *enforceable encryption* — software that uses encryption to obfuscate critical data residing on the device. As noted in Chapter 9, extensible memory on the devices, including removable storage, makes the loss of the device quite dangerous if it contains any sensitive data. One way to mitigate this loss is to encrypt the data on those memory cards; then, if the device is lost or stolen, unauthorized users can't use a card reader to access the memory card's data. For the same reason, the use of strong authentication techniques should be mandatory for on-board memory as well. The following sections cover the various types of enforceable encryption you can use to secure your organization's devices.

Encrypting all outbound and inbound communication

If your goal is to protect the whole data ecosystem, you need mandatory encryption of all outbound and inbound communication — that is, all messages to and from the device. On the face of it, this is no different from the policies that are imposed on the users of laptops and desktops that connect to your network via VPNs (virtual private networks).

There is, however, one important way that mobile device encryption must differ from typical laptop encryption: Your polices must address the ever-expanding set of customized applications that mobile device users constantly download and experiment with.

Although a majority of these applications have nothing to do with you — because they don't access any enterprise content — they do pose a problem for a potential *encrypt all* policy. You'd have to transport all that non-enterprise application data, dragging it into the enterprise only to redirect it back to the Internet, as illustrated in Figure 10-6.

Encrypting only enterprise traffic

The obvious alternative to this approach is to discern enterprise applications from non-enterprise applications and intelligently encrypt only the traffic destined for the enterprise. That's a win-win for everyone, right? Well, not exactly. The solution requires a smart agent to reside on the mobile device and make the decision of what traffic to encrypt and what to let fly, as shown in Figure 10-7.

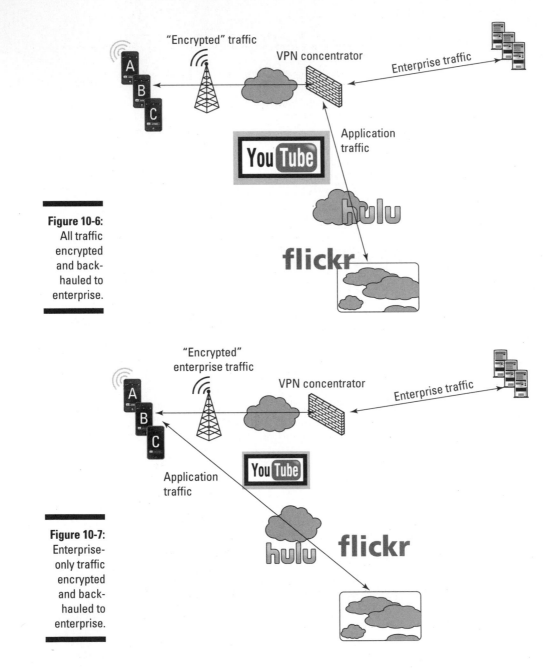

Figure 10-6:
All traffic
encrypted
and back-
hauled to
enterprise.

Figure 10-7:
Enterprise-
only traffic
encrypted
and back-
hauled to
enterprise.

Voice communication exploits

Now, this may be overkill for the average enterprise, but . . . with the widespread adoption of smartphones comes a tendency to use them to conduct mission-critical business. That makes the smartphone a very juicy target for all the vandals out there. And while there haven't been widespread exploits against voice communication so far, the day can't be far away when this critical conduit becomes a frequent target for attackers. In fact, as recently as January 2011, a European researcher showed how some bugs he discovered in the baseband chipset firmware of iPhone and Android smartphones could be exploited to ultimately take control of these devices. (Let's just hope the wrong people weren't taking notes.)

More uneasy news: Ralf-Philipp Weinmann, a researcher at the University of Luxembourg, demonstrated an exploit he created that turns on the auto-answer feature of a smartphone and then uses it as a remote listening device.

With mandatory two-factor encryption on a smartphone, the auto-answer feature would work only when calls from trusted third parties were answered. And then the communication would be encrypted, so the hacker couldn't decipher the conversation anyway.

You have to depend on the device manufacturer and the OS vendor to supply the supported encryption algorithms, but at least it's a mature technology. Most of these new devices offer pretty comprehensive feature support, so it shouldn't be a problem area.

Using carrier-provided voice encryption

Another type of encryption starting to become prevalent is *carrier-provided voice encryption*. Yes, I know what you are thinking: "Isn't there some form of encryption already provided by the radio technologies in the network?" Good question — and the answer is yes. Although such embedded encryption has been robust for years, lately it's starting to show signs of weakness. In fact, in 2009, Karsten Nohl cracked the famed A5/1 GSM encryption. But what took the cake was the modest level of technology it took to crack the encryption scheme — RF equipment, patience, and a $500 laptop!

To put this event in context, A5/1 is the most common encryption technology. It's used by over 80 percent of subscribers worldwide.

This is the point at which the additional encryption provided by carriers starts to make sense, specifically for your most critical users (chief-level officers, sales heads, and so on). With sensitive data at stake, adopting voice encryption is not too outlandish.

The most common way to implement voice encryption is to use a carrier-provided two-factor encryption solution:

✔ Each smartphone gets a hardened, self-contained crypto engine inserted into its microSD slot.

  • The phone gets the strength of additional hardware authentication.

  • Members of a defined group of trusted users can exchange encrypted calls.

  • You can manage this capability over the air.

✔ Users can easily place and receive encrypted calls by integrating with the mobile phone's standard operation and address book.

  • This security function is now on-demand.

  • Mutual authentication and end-to-end encryption make a high-security call mode possible.

Case Study: AcmeGizmo Endpoint Security Deployment

Back at AcmeGizmo, Inc., Ivan the administrator has run into a serious issue. One of the executives downloaded some applications to his Android device, and as it turns out, one of those applications was a phishing application made to look like the Customer Relationship Management (CRM) application that AcmeGizmo uses. As a result, the bad guys who posted the app got their hands on the executive's login credentials.

Luckily, Ivan and his team changed the password before anyone could log in with the stolen credentials and get access to the AcmeGizmo customer database, but this issue was a huge wake-up call that spotlighted the importance of endpoint security for mobile devices. The CEO of AcmeGizmo demanded that all mobile phones with access to corporate information be equipped with endpoint security software — immediately, and securely ever after.

Ivan started his search by re-evaluating the AcmeGizmo mobile device security policy. While developing this policy, Ivan had included a section on endpoint security. Sadly, he hadn't yet gotten around to deploying a solution to actually *secure* those endpoint devices, which was a big mistake, given the amount of attention this recent incident got. After talking to IT admins at a few other enterprises and evaluating some vendor solutions, Ivan decided to equip AcmeGizmo mobile devices with these basic security features:

- ✔ Antivirus
- ✔ Personal Firewall
- ✔ Antispam
- ✔ Encryption

That finally got some good, solid, basic endpoint security to where it was needed. Ivan slept better afterward. But he still jumps a little when his smartphone rings.

Endpoint security

Several of these technologies are covered in the same solution that Ivan selected for AcmeGizmo's VPN access: Junos Pulse. That solution provides the antivirus, personal firewall, and antispam capabilities that Ivan's corporate security policy requires, making it easy for him to simply enable the functionality inside his network. And since the smartphones are already running Junos Pulse for VPN, there's no need to distribute additional software or try to get the employees to download it. In addition, the tie-in to the SSL VPN allowed Ivan to implement policies that prohibit end users from connecting to the corporate network if endpoint security applications are not installed and running on their smartphones or tablet devices.

Had Ivan already installed and enabled this solution, the phishing application that found its way into the executive's smartphone would have been neutralized and removed before the executive could even attempt to input the username and password that the hacker wanted to steal. This would have saved a lot of trouble.

Ivan decided to implement the antispam functionality (in addition to the personal firewall and antivirus functionality) primarily because he anticipates that voice and SMS/MMS spam will become a problem for AcmeGizmo employees in the future. He hasn't yet heard overwhelming demand for such a solution, but he's (understandably) vigilant: There have been a few complaints, and he's been reading in trade journals that these types of spam are becoming more prevalent. The antispam functionality provides additional coverage, supplementing the antispam solution that protects AcmeGizmo's Microsoft Exchange Server (which processes the corporate e-mail).

Device encryption

Ivan has also decided to implement encryption on AcmeGizmo mobile devices. (Smart guy, Ivan.) Since sensitive corporate data is stored not only on the primary smartphone disc but also on removable media such as SD cards, Ivan created a policy that covered both those hardware components.

He surveyed the various devices and operating systems in use across the company, and he quickly found that some of those devices had native or built-in encryption capabilities, while others required additional third-party software in order to encrypt.

On platforms where encryption was native, Ivan configured the AcmeGizmo Mobile Device Management (MDM) solution to ensure that each device complied with the encryption policy. On other platforms, Ivan had to purchase a third-party software solution to enable device encryption. As with the native devices, Ivan made sure that both the on-device disc and any removable media were protected by the encryption product.

Flash forward

Several weeks later, Alvin from Accounting downloaded a malicious application from a shady website to his smartphone. Fortunately, the antivirus solution detected that application, saving Alvin (and Ivan) from the embarrassment of another malware incident on one of the company smartphones. This was very fortunate because (as is the case with more smartphones all the time) the AcmeGizmo smartphones contained sensitive corporate data. Alvin's device, for example, contained all the company financial data from the recent quarter, because he wanted to review some of it while on vacation with his family.

Not only is the device protected from malware, but the encryption — along with the capability to wipe the device remotely if Alvin were to lose it — also has Ivan sleeping much better at night. Especially when he can turn off his smartphone.

Chapter 11

Protecting Against Loss and Theft

*1*dentity thieves look for easy pickings on smartphones — devices that can be unlocked without a password or apps that have stored passwords. These types of weaknesses can let a thief access critical information easily on a device and steal personal and corporate contacts, data, and other information without breaking a sweat.

This chapter covers the services available to protect users' personal and corporate data from thieves. If a user's phone has been lost or stolen, she can take actions to prevent a thief from stealing the data on that phone. For example, she can first try to find the phone using GPS. If she's unable to locate it using GPS, she can remotely lock the device so a thief can't unlock it and access the data on it. Other options include remotely setting off a loud alarm that the thief can't turn off, obviously attracting the attention to the person carrying the stolen phone. These services can alleviate a lot of anxiety for users (and you!) in the case of a lost or stolen phone.

In this chapter, we cover both consumer-grade and enterprise-class solutions to protect mobile devices from loss and theft. Additionally, we look at how a corporate loss and theft protection plan should be deployed across an enterprise.

Taking Precautions before Loss or Theft

As an administrator managing mobility for an enterprise environment, you want to enforce certain policies for all personal devices being used at work, especially if your IT policy allows personal devices to access network resources.

Here are some precautions you should advise users to take to prevent losing vital data on a smartphone if it's ever lost or stolen:

- ✔ **Add a device password.** Every smartphone needs to be protected by a password. Not setting one is simply too risky, because a thief can easily access all the phone's information without having to guess the password. Also, users need to make sure that the phone is set to lock automatically after a certain duration of inactivity.

- ✔ **Back up often.** The contents of the device must be backed up regularly, including photos, contacts, and videos. In the unfortunate circumstance that the device is lost, the phone's contents may need to be wiped remotely to prevent a thief from accessing the information. As an IT administrator, you can choose to either deploy a corporate backup and restore system or advise users to manage their own backups and restoration. Having backups is particularly useful if a user loses his device and the device needs to be remotely wiped. When the user gets a new device to replace the lost one, the backed up data can easily be restored to the new device.

  Chapter 12 describes backups and restorations in more detail, and in Chapter 15, we discuss several commercial enterprise solutions for mobile backups.

- ✔ **Store the device's IMEI number.** The device's IMEI, or International Mobile Equipment Identity, is a 15-digit number that uniquely identifies it. Carriers use this number to identify and track the device. Ask device owners to locate the IMEI number using appropriate techniques for their phones and store it in a safe place. Different devices have different techniques for locating the IMEI number. For example, an iPhone's IMEI number can easily be retrieved by using iTunes.

- ✔ **Deploy antitheft services.** Several carriers offer insurance or other antitheft services for smartphones, including the ability to remotely lock, locate, or wipe devices. Device owners can purchase many such services from the carrier directly. Some device vendors also offer these services for their specific device types. For example, HTC offers HTC Sense with the ability to remotely lock or wipe devices. Apple's MobileMe is another example of a service that device owners can deploy themselves.

  Enterprise-grade services are available from vendors like Good Technology, McAfee, Juniper, and many others. The Junos Pulse solution from Juniper includes corporate remote access along with mobile antitheft and security services. Depending on the scale and nature of your need, you'll find an appropriate solution out there. Chapter 15 discusses such solutions in more detail, so be sure to check it out.

The difference between a personal (consumer) solution and an enterprise-ready solution is that the latter lets you — the IT administrator — enforce mobile policies from a central management console. A personal solution relies solely on users doing the "right thing" in terms of setting passcode policies or remotely wiping the device if it's been lost or stolen.

Educating Users about Securing Data on a Lost Phone

Despite your (and your users') best attempts to prevent the loss or theft of a device, it does happen. If a user loses a device accidentally, there are ways to minimize the potential damage it could result in. Here's a list of actions to ask users to take to prevent access to the data on their lost phone:

- Inform your mobile carrier that you've misplaced your smartphone or tablet (if it's a 3G-enabled device).
- Locate the device using GPS.
- Remotely lock the device so the thief can't unlock it.
- Remotely set off a loud alarm on the device.
- Remotely wipe the device's contents and reset it to the factory default settings.
- Detect SIM card changes on the device.

How users protect a mobile device like a smartphone or tablet from loss and theft depends on the type of device and the operating system it runs. Nearly every mobile operating system has its own way of providing loss and theft protection. If your organization supports personal devices within the workplace, the choices of devices and options to enforce protection from loss and theft can be intimidating.

Consumer-grade, as well as enterprise-class, solutions are available to protect mobile devices. In the following sections, we focus on consumer-grade services. If you are an administrator looking to protect thousands of mobile devices at work, an enterprise-grade solution is a better option. For more information, check out "Exploring Enterprise-Grade Solutions for Various Platforms" later in this chapter.

In the sections that follow, we highlight some of the leading platforms and devices and talk about the availability of loss and theft protection services for your users. If you're not deploying an enterprise-grade solution, this is what you should recommend to your users.

Protecting personal Apple iOS devices

Apple offers a sophisticated solution that users can employ to protect personal iOS devices such as iPhones, iPads, and iPod touch devices. Apple's Find My iPhone service is a feature in MobileMe, and it's also available for free for iPhone 4 users and iPad and iPod touch devices.

At the time of this book's writing, Apple was transitioning its MobileMe product and service to its new iCloud product, and Apple was still a little nebulous about all the details. Depending on when you're reading this book, the product could still be MobileMe, or it could have its new name and feature set. Rather than go off of rumors about a unreleased but announced product, we just refer to MobileMe throughout this chapter. If the URLs provided change, we're fairly certain that Apple will redirect you to the proper pages. So MobileMe/iCloud, here we go.

MobileMe requires users to register at www.apple.com/mobileme with an Apple ID. After they've registered their device (per the instructions described by Apple at the preceding web page), they can locate, lock, or wipe devices remotely. The following actions available in MobileMe are key to protecting Apple iOS devices:

- ✔ **Locate lost devices using GPS.** Locating a lost device using GPS is perhaps the first step that users might take if they lose their smartphone. Apple's MobileMe Find My iPhone service provides this feature as long as the device is registered for it. Figure 11-1 shows what users see on their computer screen when searching for an iPhone with MobileMe.

- ✔ **Remotely lock or wipe the device.** Another option is to lock the device remotely so the person who has the phone can't retrieve the contents from it. This is especially critical if the user hasn't set up a password for the phone.

  Apple's MobileMe provides this option. Users need to log in to MobileMe via a web browser. Once logged in, they simply select the action to remotely lock the device. (See Figure 11-1.) The device remains locked until the user chooses to unlock it again from the same web page.

  If the user feels that the device is indeed lost, the best course of action may be to simply wipe its contents to prevent any data from falling into the wrong hands. In that case, the same service allows the user to remotely wipe the device, thereby resetting it to the factory default state.

- ✔ **Remotely play a sound or message.** If the user can't find the phone, he might want to play a sound on the lost device to attract attention to it, or simply display a message on it, something like "I've lost my iPhone; if you find it, please call me at 555-555-5555." Both options are available via MobileMe. The user needs to log in to the MobileMe service on a computer and set a message to be displayed or select a sound to be played on the device. (See Figure 11-2.)

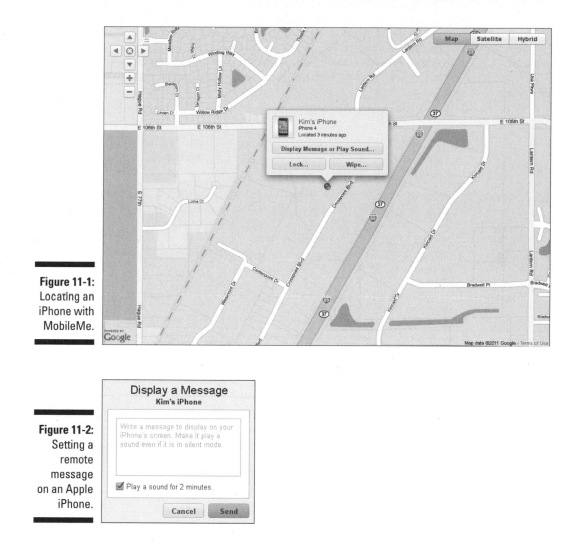

Figure 11-1:
Locating an
iPhone with
MobileMe.

Figure 11-2:
Setting a
remote
message
on an Apple
iPhone.

Protecting personal Symbian devices

Devices running Symbian include Nokia's smartphones, such as the N-series and E-series devices. These smartphones run the Symbian mobile operating system. Users can take the following actions to protect their personal devices from loss or theft:

✔ **Locate the lost device using GPS.** When a Symbian phone is first purchased, the user should inquire with the wireless carrier whether a GPS locating service is available. Some carriers offer services for various types of phones for a nominal monthly or annual charge.

Alternatively, vendors like McAfee WaveSecure or Kaspersky offer online services that users can buy to help locate Nokia devices. These services largely work in the same way as MobileMe for Apple iOS devices: They provide a web-based interface that users can log in to, in order to locate the missing phone. These services typically cost $20–30 per year and can help alleviate panic when a phone is lost.

✔ **Remotely lock or wipe the device.** If the user is unable to find the device using GPS, many Symbian devices have built-in capabilities to remotely lock the device or remotely wipe it and reset it to factory default settings. The device can also be wiped automatically if a wrong passcode is entered too many times. These settings are built into the device itself and can be activated easily.

To set up remote locking and wiping before the device is lost or stolen, users must do these two things: Enable remote locking and set a wipe message. The steps are somewhat different depending on which device the users own, so they may need check the device's documentation for clarification. If it's an N97 phone, for example, users should go to Settings⇨Phone⇨Phone Mgmt.⇨Security Settings⇨Phone and SIM card. Users then see a Remote Phone Locking option that can be either enabled or disabled, as shown in Figure 11-3.

Enabling that setting activates this feature, which works by sending an SMS message from another phone to the lost phone with a message — like "wipephone" — that the user has set on the phone. Once the setting is activated, the user can remotely lock or wipe the device by sending an SMS message with the text he configured on the device.

Wiping the device removes all personal contents from the device, leaving it in a factory-default state. This makes regular backups all the more important. Regular backups can save the user in such situations by enabling him to restore to a previously saved configuration.

✔ **Remotely set off an alarm.** If a user can't find his lost Symbian device, he may want to remotely set off alarms on it. Many Symbian devices have built-in capabilities to remotely wipe the device when a SIM card change is detected. This prevents a thief from simply removing the phone's SIM card and replacing it with a different one. When such a change is attempted, the device wipes its own contents, thereby preventing any personal data from being visible to the thief.

Even if the person is left with a stolen Nokia device with factory default settings, no personal data is compromised. At the end of the day, that's what matters to prevent identity fraud from occurring. Users can check the device's documentation for information on how to set a remote alarm.

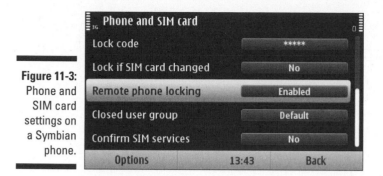

Figure 11-3:
Phone and
SIM card
settings on
a Symbian
phone.

Protecting personal Android devices

Google's Android operating system powers a number of smartphones and other devices from various handset manufacturers. Vendors such as Motorola, Samsung, and HTC make several devices that run the Android operating system. The user's ability to remotely control Android devices largely depends on the type of Android device. We show you the options available to protect some of the leading Android devices from loss and theft.

If users are looking to protect their personal Android smartphones or tablets, they have several options to remotely lock, wipe, or locate their devices:

✓ **Locate the lost device using GPS.** As soon users discover their device is missing, they may want to try to locate it. They need a web-based interface that they can log in to, for locating the lost device using GPS.

Depending on your device's make and model number, the vendor itself may provide services to remotely locate, lock, or wipe devices. Services are available to device owners from vendors like HTC, Samsung, and Google to manage these actions.

✓ **Remotely lock or wipe the device, or set off an alarm.** If users are unable to locate a lost Android device using GPS, they can remotely lock or wipe it. They can also set off a remote alarm on the Android device to attract attention to it.

These actions are also typically taken from a web page by the device owner.

Recommend that users shop for loss and theft protection services from their carriers. Several carriers offer such services that may be competitively priced in comparison with buying them directly from a vendor.

Protecting personal Windows Mobile and Windows Phone 7 Devices

Microsoft's Windows-based mobile operating system comes in two basic flavors. One is the older Windows Mobile operating system that runs on several phones from vendors such as HTC and Sony Ericsson; the other is the relatively newer Windows Phone 7. Both are vastly different operating systems with different sets of supported features.

Microsoft offers a service called My Phone for Windows Mobile and Windows Phone 7 devices, with support for loss and theft protection features, as well as the ability to sync photos, music, and other data from the phone to a computer.

Microsoft's My Phone service (http://myphone.microsoft.com) is available in two options. One is a free option with limited features, and the other is a premium option with fully supported features.

If the user's phone is lost or stolen, he can use the following My Phone features to protect the data:

✔ **Locate the lost device by using GPS.** After the user has signed up for Microsoft's My Phone service, he can locate a lost or misplaced device by logging into the service using a web browser from any computer.

Note that the free version of Microsoft's My Phone service provides the location of a device when the last sync operation was performed. This clearly isn't the same as locating the actual device when it's lost or stolen. For example, if the user last performed a sync a month ago, it will show the location where the sync was performed, not the current location of the phone.

✔ **Remotely lock or wipe the device, or set off an alarm.** If the user is unable to locate the lost device using GPS, he should attempt to remotely lock the phone or even wipe its contents clean. The My Phone service provides these services for a fee.

Remotely locking the device with My Phone involves using the web-based My Phone interface and setting a passcode to lock the device. The device owner can also use the My Phone web interface to issue a wipe command on the device if he's fairly certain it has been lost or stolen.

Other solutions available commercially from other vendors include similar web-based interfaces to issue remote lock or wipe commands.

Protecting personal Blackberry devices

RIM offers a BlackBerry Protect app through the BlackBerry App World. This app allows users to protect their personal BlackBerry devices from loss and theft by giving them the ability to remotely lock, wipe, or locate their devices. The app also enables users to back up to a remote BlackBerry server and to restore the backup to a new BlackBerry device if the old one is lost.

BlackBerry Protect is available for users to download for free from www. blackberry.com/protect or from the App World on the device. Users can configure device backups at regular intervals of time and specify what types of data should be backed up (contacts, text messages, calendar, memos, tasks, and bookmarks).

From the BlackBerry Protect website, users can log in and take actions on their device, such as remotely locking or wiping the device.

Exploring Enterprise-Grade Solutions for Various Platforms

If you want to deploy a corporate solution to protect your employees' mobile devices, you need what is generally referred to as a mobile device management (MDM) solution, which provides the remote management, provisioning, and configuration capabilities that enable you (the administrator) to take actions such as remotely locking or wiping devices.

Enterprise solutions typically include a web-based management interface that you can access from any web browser on a PC or Mac. The management interface allows you to manage a large number of devices, including remotely locking, wiping, or locating lost or stolen devices.

An enterprise MDM solution comprises a combination of mobile software apps deployed on the devices themselves, and a management console for administrators. Most MDM solutions offer the management server as an in-house appliance — SaaS (software as a service) — or a virtual appliance.

Enterprise-grade solutions for Apple iOS

If you're looking for an enterprise-class solution scaling for thousands of iOS devices, look for solutions from vendors like Juniper Networks or MobileIron, which offer some of the more attractive and effective loss and theft protection for iOS devices. Juniper's Junos Pulse solution and MobileIron's Virtual

Smartphone Platform are products available for enterprises to protect their employees' lost or stolen devices.

Enterprise-grade solutions for Symbian

Solutions from Juniper Networks and WaveSecure (now McAfee) are among the more attractive and effective solutions for loss and theft protection on Symbian phones. Juniper's Junos Pulse solution provides a good combination of enterprise features, including SSL VPN and loss and theft protection. Junos Pulse also includes a web-based management console that provides for easy administration of remote devices, scaling to thousands of devices.

Enterprise-grade solutions provide you with the ability to set passcode policies, locate or wipe stolen devices, or restore previous backups to devices. You can enforce and own actions that otherwise would be the user's responsibility.

Enterprise-grade solutions for Android

One of the challenges of managing Android devices within an enterprise is the diversity of those devices. Android devices can range from smartphones built by Motorola, Samsung, or HTC to tablet devices from the same or other vendors. Employees are bringing all types of Android devices to work, thereby presenting the challenge to IT administrators to centrally manage all of them.

Several vendors offer competitive solutions for managing and securing Android devices. Solutions from Juniper Networks, WaveSecure (now McAfee), and Lookout are among the more attractive ones to protect Android devices from loss and theft.

When shopping for an Android solution, look for the diversity of devices and platforms supported by the vendor's solution. Look for vendors that support all recent Android versions, or else you'll be unable to support employees that have devices running certain versions.

Enterprise-grade solutions for Windows Mobile and Windows Phone 7

If you're looking for enterprise-class solutions to provide loss and theft protection on Windows Mobile phones, several vendors provide such solutions. Solutions from Juniper Networks and Lookout are among the more attractive ones providing loss and theft protection for Windows Mobile smartphones.

At the time of writing this book, because Windows Phone 7 is still relatively new, there isn't an enterprise-class mobile security solution available in the market for Windows Phone 7. The only solutions available for Microsoft's mobile operating system are for Windows Mobile–based smartphones.

Enterprise-grade solutions for Blackberry devices

Devices running the BlackBerry operating system from RIM are typically protected and managed using the BlackBerry Enterprise Server (BES). BlackBerry devices are widely used within corporate environments and can be managed using the BES. The BlackBerry Enterprise Server can deploy apps to the managed devices and also send remote commands to the devices, such as to lock or wipe them.

For a corporate deployment, other solutions are also available to provide loss and theft protection to BlackBerry devices. Vendors such as Juniper Networks, McAfee, and Lookout provide attractive solutions that enable you to remotely lock, wipe, or locate lost or stolen BlackBerry devices.

These solutions protect Blackberry devices against loss and theft by allowing you to take the following actions:

- ✔ Remotely locate the device.
- ✔ Remotely lock or wipe the device.
- ✔ Remotely set off an alarm on the device.
- ✔ Detect SIM card changes and automatically wipe upon detecting a SIM change.

Deploying Enterprise-Wide Loss and Theft Protection

If you need to deploy loss and theft protection on hundreds or thousands of devices across your enterprise, you need to identify the platforms you want to support in your enterprise and then shortlist the vendors whose solutions you'd like to deploy.

Here are the specific steps you need to take to start planning your deployment:

1. **Determine the types and platforms used in the enterprise.**

   Find out what types of devices and platforms you need to support across all your users in the enterprise. Can you limit yourself to supporting only

Apple's iPhones and iPads, or will you need to support Android devices as well? And what about BlackBerry devices, or the users who've bought themselves the shiny new Windows Phone 7 devices?

2. **Shortlist the solutions feasible to protect each type of platform.**

From your analysis of all the individual platforms, make a list of the available vendors and solutions to provide loss and theft protection for all the platforms you need to support.

Table 11-1 is an example of what your list might look like.

Table 11-1	Sample Loss and Protection Worksheet		
Platform I Need to Support	*Vendor #1*	*Vendor #2*	*Vendor #3*
Apple iOS	Juniper	MobileIron	
Android	Lookout	Juniper	McAfee (WaveSecure)
BlackBerry OS	Juniper	Lookout	McAfee (WaveSecure)
Symbian	McAfee (WaveSecure)	Juniper	
Windows Mobile	Juniper	McAfee (WaveSecure)	

3. **Evaluate free trials of the shortlisted solutions.**

After you've identified the leading solutions of loss and theft protection for the platforms you need, explore free trials. Many vendors offer limited free trial durations (such as for 30 or 60 days), which will help you evaluate those solutions.

In the rapidly changing world of smartphones and other portable devices, look for vendors whose solutions aren't restricted to just one or two mobile platforms. Because employees are likely to buy more than just one or two types of devices, it's ideal to roll out a solution that will address everyone's needs and provide a comprehensive loss and theft solution.

Case Study: AcmeGizmo's Lost or Stolen Device Recovery

Returning to our ongoing AcmeGizmo case study, recall that Ivan, the IT manager, had fully deployed a smartphone solution for several hundred employees of AcmeGizmo. This solution included a mobile security solution that protects the mobile devices and the data on them, as well as a VPN solution, which authenticates users and encrypts any data transiting between the mobile device and the AcmeGizmo network.

Ivan was at his desk one afternoon when he received a frantic call from Ed in Engineering. Ed had been sitting in the food court at the mall enjoying lunch when he reached into his pocket to check the e-mail on his new Android phone, only to realize that it was not there. He checked his jacket pockets and shopping bags to no avail. Ed's biggest concern was the fact that his phone contained the latest revision of the next-generation widget design that he and his team had been working on, despite a corporate policy against storing such data permanently on mobile devices. If this information were to get into the hands of AcmeGizmo's competitors, their groundbreaking design would be compromised.

The first thing that Ivan did was log into the Junos Pulse Mobile Security Gateway and lock the device. Figure 11-4 shows the two commands (Handset GPS Location and Handset Lock) that Ivan sent to Ed's mobile device. Fortunately, Ivan was able to successfully lock Ed's device and track it via the GPS feature on the phone. Unfortunately, it appeared as though the device were in someone's vehicle heading northbound away from the mall on the freeway.

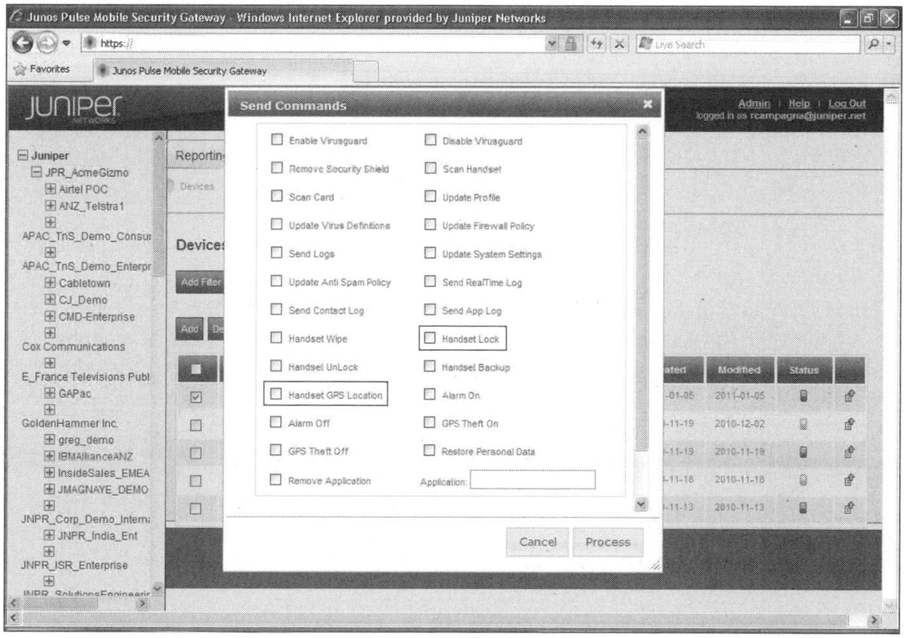

Figure 11-4: Remotely locking and locating Ed's lost mobile device.

Ivan's next step was to turn on GPS Theft mode, which constantly updates the Mobile Security Gateway with the device's location and emits an audible "scream" or alarm on the device. Ivan was able to emit the alarm even though Ed had left the device in silent mode.

Ivan also initiated a wipe of the handset, as shown in Figure 11-5. Despite not being able to confirm that the thief actually wanted access to sensitive corporate data, best practices dictate that the handset be wiped in this case, to avoid that data getting into the wrong hands.

From that point, Ivan realized that his investment in this security infrastructure had paid off — the device had been wiped. Because the location tracking was enabled, Ivan was also able to report the theft to the local police department so that they could recover the device.

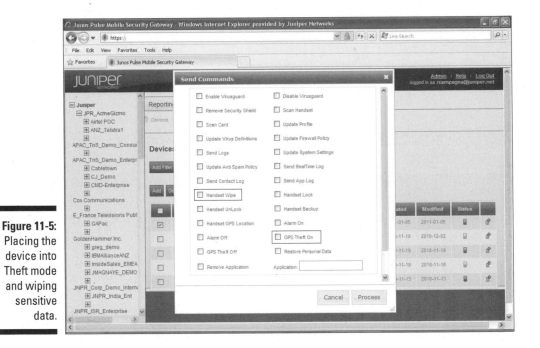

Figure 11-5: Placing the device into Theft mode and wiping sensitive data.

Chapter 12

Educating Users about Backing Up Data

In This Chapter

▶ Backing up data from a mobile device

▶ Restoring data from a mobile device backup

▶ Transferring data from one mobile device to another

*J*ust as we detail in Chapter 11, there are practical things you must tell your employees about mobile device security. In this chapter, we give you information that you can readily pass along to users, written at their level. Check the facts because some of the procedures may have changed since the writing of this book, but use the material in this chapter to show users what to do to back up, restore, and transfer data.

Smartphones, tablets, iPads, and other portable devices that work along similar lines are increasingly vital to our daily lives. On these devices reside plenty of vital personal and corporate data, including e-mails, SMS messages, contacts, call logs, photos, and videos. Anyone who has lost or misplaced a smartphone can vouch for the fact that losing such a device causes a lot of angst, especially if the device's contents aren't backed up to a computer or other storage device.

This chapter looks at the types of data that users need to back up from their mobile devices and smartphones. We also look at the tools available to back up and restore data for different types of mobile devices.

Backing Up Data from Smartphones

As smartphones and tablets steadily gain adoption, the amount of valuable data on them is also growing. Employees bring many of these devices into the corporate workplace and use them to access e-mail and other applications. As an administrator managing mobility policies for your corporate users, you may need to take actions such as remotely locking or wiping a device when it's lost. In such circumstances, when the user gets a new device, it becomes a simple matter of restoring the previously backed up configuration to the new device. Having backups not only benefits the users who can get their data onto their new devices easily but also enables you to take actions such as wiping the device, knowing that the contents are securely backed up.

It's critical for device owners to back up their devices periodically to avoid losing data accidentally. As an administrator, you must encourage users to back up their devices often and explain how to do it. This chapter describes in detail the steps that device owners can follow to regularly back up all the contents or specific contents of their devices.

Mobile devices such as smartphones and tablets have the following types of data that need to be backed up:

✔ Personal files, including videos and photos

✔ Call log and contact information

✔ Apps and app settings

✔ SMS messages

✔ E-mail and calendar information

✔ Phone settings (backgrounds and other customized configurations)

Though almost every smartphone has the ability to store the preceding types of data, the mechanisms to back up or restore data differ drastically from one device type to another.

Most smartphones rely on the technique of backing up data when the device is physically connected to a computer. Desktop applications such as BlackBerry Desktop Manager, Apple iTunes, and Nokia PC Suite can be installed on the user's computer. These applications can back up some or all of the preceding types of data from the device to the computer. They can also restore data from the computer back to the device. Encourage users to back up their smartphones or mobile devices often by connecting those devices to their computers.

Some commercial software applications can back up the smartphone's contents periodically without requiring that the device be connected to a computer. These applications are especially useful if you want to deploy a corporate-wide solution for all your company users. We look at such applications in Chapter 15 as well as in the section "Exploring Corporate Solutions for Backup and Restore," later in this chapter.

Instructing Users on Backing Up Their Devices

In the following sections, we give you examples of the type of hands-on instructions that you'll want to pass along to your users for backing up their devices.

Backing up iPhones and iPads

Apple provides the iTunes software application that can be installed on Windows PCs and Mac computers. The iTunes application (www.apple.com/itunes) provides a simple user interface to control backups as well as to restore data from the computer back to your device, if needed.

To back up an iPhone and iPad to your computer, connect your device to your computer and sync it with iTunes. iTunes makes a backup each time you sync the device to your computer.

Only one copy of the backup is maintained, so even if you back up more than once, only the most recent copy is maintained on the computer.

iTunes backs up most device settings, including call history, apps, contacts, e-mail, Wi-Fi, and VPN settings, and several other pieces of information. You can customize the actual data that's backed up by selecting the appropriate options on the iTunes application. This is useful if you want to back up specific data, instead of backing up all the contents on the device. To customize the data being synced between your iPhone or iPad and your computer, follow these steps:

1. **Connect the iPhone or iPad to the computer that has iTunes installed.**

   iTunes should start up automatically when it detects the iPhone or iPad connected to the computer.

2. **In iTunes, under Devices, select the device being synced.**

3. **In the right-hand pane, select the application type whose settings you want to customize.**

   The following application types should be visible:

   - Info
   - Apps
   - Music
   - Movies

- TV Shows
- Podcasts
- Books
- Photos

4. **Make changes to the application type as needed and then click Apply (at the bottom-right) to confirm your changes. (Otherwise, the changes won't take effect.)**

5. **Repeat Steps 3 and 4 as desired.**

You can configure your backed-up data to be encrypted on your computer, which prevents the files from being opened and read by making them unreadable without the passcode. On the Summary tab, select the Encrypt iPhone Backup check box and enter a password. Be sure to remember the passcode, though, or you won't be able to restore to the device.

Backing up Android devices

Google uses its secure servers on the Internet to store your device's configuration.

Each Google Android phone uses the Google account associated with the device to sync the phone's contacts, apps, and settings. If you have a Gmail account registered on the device, the contacts, settings, and apps stored on the device are synced automatically with that Google account. For example, as soon as you add a contact in your phone, the information is automatically synced to Google servers. The information you sync to the Google servers (said to be *in the cloud*) is always available to sync with any Android device you choose.

Automatic syncing with Android

Users with mobile devices running the Google Android operating system back up their devices to Google servers in the cloud. This should not sound alarming — at least any more than the fact that the users may already be using other services in the cloud, like Gmail. The contents backed up to the Google servers doesn't include sensitive application data.

Keep in mind that Google's automatic syncing only helps back up contacts, phone settings, and apps installed on the device. It doesn't back up SMS messages, call logs, and other information residing on the phone. Several applications for backing up such data are available for purchase from the Android market. Popular examples include the Sprite Mobile Backup and Astro apps. If you're looking to deploy an enterprise solution to perform automatic Backups and Restores, refer to Chapter 15 for a list of commercial solutions available.

Backing up BlackBerry devices

You can back up your BlackBerry device by using the BlackBerry Desktop Manager, which is available as a free download from the BlackBerry website (http://us.blackberry.com/apps-software/desktop). With the BlackBerry Desktop Manager, you can back up all your key information, including personal files, call logs, contacts, apps, SMS messages, e-mail and phone settings, and calendar information.

To back up data or apps from your BlackBerry to your computer, follow these steps:

1. **Connect the BlackBerry to your computer.**

2. **Start the BlackBerry Desktop Manager application. Wait for it to detect your connected BlackBerry.**

3. **Select Backup and Restore on the Main Menu.**

4. **Select Backup and enter a filename to indicate where to save the backed-up contents.**

   A progress bar appears, indicating the status of the ongoing backup.

You can configure the BlackBerry Desktop Manager to back up all contents of the BlackBerry smartphone or only the contents you configure. To select what types of data you want to back up to your computer, follow these instructions:

1. **When the BlackBerry Desktop Manager has detected the BlackBerry, select Backup and Restore on the Main Menu.**

2. **Select Advanced on the next screen.**

   A number of device databases are listed. Each refers to a category of data that can be backed up. These include the Address book, E-mail Settings, Application Data, and Call Log information.

3. **Select the appropriate data to be backed up.**

4. **Enter a filename to indicate where to save the backed-up file and then select Save.**

   The backup process begins.

This is a great option if all you need to do is back up a specific type of data. For example, if you lose your e-mail settings or maybe a few contacts in your address book, you can simply update those items without affecting the rest of your phone.

BlackBerry Enterprise Server (BES)

The BlackBerry Enterprise Server (BES) is an enterprise-grade solution that provides limited backup services for device settings and configurations. Enterprise administrators can periodically back up thousands of corporate BlackBerry devices automatically using the BlackBerry Enterprise Server. The BES backs up device settings, bookmarks, and other configuration data.

As an enterprise administrator, you should recommend that users back up their devices regularly using the BlackBerry Desktop Manager because the BES doesn't back up third-party applications, pictures, or mail messages stored on the device. To ensure that all data is backed up, the device owners are best advised to back up their own devices regularly.

With the BlackBerry Desktop Manager, you can configure backups to occur periodically or manually. If you configure periodic backups, you must connect your BlackBerry device regularly to your computer so the backup can be completed. The device must be connected to your computer to be backed up successfully.

Backing up Nokia devices

For Nokia devices such as the E-series or N-series phones, you can use the Nokia PC Suite to back up the following types of data from your phone to a computer:

- ✔ User files from the phone (includes videos, music, and other user documents)
- ✔ Contacts and Calendar
- ✔ Notes and Messages
- ✔ Settings
- ✔ Bookmarks

The Nokia PC Suite is currently available only for Windows PCs. It's available for download at

```
http://europe.nokia.com/support/download-software/pc-suites
```

Not all Nokia devices are supported by the Nokia PC Suite. Be sure to use the PC Suite application to check the compatibility of your specific Nokia device. You can do so at the following location:

```
http://europe.nokia.com/support/download-software/pc-suites/compatibility-and-
                              download
```

To back up data or apps from a Nokia device to a computer, simply follow these instructions:

1. **Connect the device to the computer.**

2. **Start the Nokia PC Suite application and select Backup.**

3. **In the next window, select Backup to back up the device.**

   A Restore option is also available to restore contents to your device.

4. **Select a suitable location to which to save the backed-up file.**

5. **Select Next to start the backup.**

If you want to back up specific types of data, such as only your photos or videos, you can customize the PC Suite to do so. When the PC Suite starts up on your Windows PC and prompts you for backup options, select the required content types to back up. You can select any or all of the supported content types, such as user files, contacts, calendar, and phone settings.

Backing up Windows Phone 7 devices

For Windows Phone 7 smartphones, you can use the Zune software from Microsoft to sync files and data between those devices and your computer. Zune enables the transfer of files such as videos, music, and pictures, as well as call logs and phone settings from a smartphone to a computer. Zune is also required to update a Windows Phone to the latest software versions released by Microsoft.

The Zune software is available as a free download from

```
www.microsoft.com/windowsphone/en-us/apps/zune-software.aspx
```

The Zune software is available only for Windows PCs at the time of writing. There is, however, a Windows Phone 7 Connector application available for Mac computers. You can download it from the Microsoft Download Center (www.microsoft.com/downloads).

The Zune software allows you to back up specific files or data selectively, such as particular videos of corporate training programs or maybe pictures of a product launch or marketing event. There's no need to sync the entire device with the computer every time.

Zune can also be configured to sync a phone and computer automatically whenever they're both connected to the same wireless network — as when, for example, your phone and PC are both connected to a company wireless network.

Automatic backups using Zune kick in only when the phone isn't in active use. For example, your phone has to be sitting idle for the automatic backup to start with the computer. If it's active — say, you're having a phone conversation or browsing the web — the backup won't run.

Zune does *not* back up any apps installed on a Windows 7 phone. For example, if you purchased an app from the Windows marketplace and customized it with your own configuration and data, it won't be backed up anywhere.

Instructing Users on Restoring Data to Their Devices

Devices typically need to be *restored*, or returned to a previous backup configuration, when they lose existing data or configuration for any reason or when they are wiped of all contents intentionally. For example, if a device OS upgrade fails, a restore can get the device back to its previously working configuration. Similarly, if a device is assumed to be lost or stolen, you can intentionally issue a remote lock or wipe command, thereby wiping the device clean of all data. Then if the device is retrieved, it needs to be restored to a previously backed up configuration.

You can restore a phone in two ways:

- ✔ Restore specific files or settings to an earlier state.
- ✔ Restore the phone to factory settings.

Hopefully, users won't frequently lose their phones or have to restore their devices to old configurations. But if they do, in most cases users can restore their devices themselves from the backed up configurations on their computers or in the cloud. So as an enterprise administrator, the steps in this section are what you would recommend to your corporate users to follow to restore data to their devices.

If you need to transfer backed-up data to a new device, see the upcoming "Instructing Users on Transferring Data to New Devices" section.

Restoring data from iPhones and iPads

You can use the iTunes software to restore data on an iPhone or iPad. The first step is to restore the device to its factory-default settings. Doing so deletes all data from the device, including contacts, videos, photos, and other user information. When the device is back to its factory-default configuration, you can restore a previously backed-up configuration to the device, thereby restoring your old data.

To restore data on Apple iPhones or iPads using a previous backup, take the following steps:

1. **Connect the device to your computer.**

2. **Select the device from the Devices option in iTunes.**

3. **Select Summary, and click Restore.**

4. **When iTunes prompts you to back up the device prior to restoring it, click Don't Back Up.**

   The device is restored to factory configuration.

5. **When the device starts up again, select a previously backed-up configuration to restore to the device.**

   Doing so will copy the old configuration to your device.

Restoring data from Android devices

For Android devices, backed-up data from your device is stored on Google's servers and associated with your Google account. When you log in using a different Android device, or restore a device to the factory-default configuration, the backed-up configuration is restored seamlessly to the device.

The procedure for resetting an Android device to factory configuration can vary from device to device. Be sure to look up the procedure in your device manual *before* you reset the device to the factory-default configuration.

When you use a different Android device or log in from a device that has been reset to factory-default configuration, Google restores the contacts, phone settings, and apps associated with your account. All those settings are restored automatically as soon as you log in to your main Google account.

The restoration of apps, phone settings, and contacts may take a few minutes; it isn't instantaneous. Be patient and wait for a few minutes for things to show up again on the device.

Call logs and SMS messages are not backed up from Android devices. So if you restore a device to factory-default configuration or buy a new Android device, these configurations cannot be easily restored.

Restoring data from BlackBerry devices

Restoring data to your BlackBerry device is nearly a mirror image of the process used to back up data: it's easy, efficient, and intuitive.

To restore data to your BlackBerry, take the following steps:

1. **Connect the device to a computer.**

2. **Start the BlackBerry Desktop Manager application. Wait for it to detect the connected BlackBerry.**

3. **Select Backup and Restore on the Main Menu.**

4. **Select Restore to start the restoration.**

   At this stage, you should remember where you stored the backed-up file. If you have more than one backed-up file, decide which configuration to back up to.

   A progress bar indicates the status of the ongoing restoration.

Restoring a device typically takes much longer than backing up a device.

Restoring data from Nokia devices

The procedure for restoring data to Nokia devices is nearly identical to the procedure for backing up data. To restore data to your Nokia devices, such as the E-series or N-series phones, follow these steps:

1. **Connect the device to your computer.**

2. **Start the Nokia PC Suite application and select Restore.**

3. **Browse to the backed-up file whose contents you want to restore.**

4. **Select Next to start the restoration.**

Restoring data from Windows Phone 7 devices

The Zune software application for Windows PCs can restore data (including photos, videos, music, and other user files) from your computer to Windows Phone 7 devices.

Such restoration doesn't, however, include apps installed on your phone. If you had customized apps downloaded from the Windows marketplace, those customizations won't be available anywhere to be restored to the phone.

Instructing Users on Transferring Data to New Devices

If a device is lost or stolen, ideally either you or the user would remotely wipe it, thereby preventing its data from falling into the hands of a thief. Then when the user has a replacement device in hand, the previously backed up contents from the earlier device can be restored to the new device.

The following sections detail the processes that your users can follow to restore contents of one device to another device running the same platform.

Transferring data between iPhones and iPads

It's easy to transfer data from an old iPhone or iPad to a new one. To transfer data from one device to another, follow these steps:

1. **Back up the old iPhone or iPad with iTunes using the instructions described in the earlier section on backups.**

2. **Connect the new iPhone or iPad to the computer.**

   iTunes prompts you, asking whether you want to restore from an old device or set up the device you're restoring as a new one.

3. **At the prompt, select the option that backs up from the old device to the new one.**

   iTunes restores contents from your old device to the newly connected device. The new iPhone or iPad restarts when data restoration is complete.

4. **You can choose to sync information selectively to the new device. Select the desired options (such as Music, Photos, or Videos), and then click Apply to continue.**

   Your selected data is transferred to your new iPad or iPhone.

Transferring data between Android devices

Google Android syncs the device by using the Google online account registered on the device. The process is similar to backing up contacts, phone settings, and apps to Google's cloud automatically.

All you need do to set up a new Android device to receive these contents from an old Android device is to log in to the same Google account using the new device. When you do so, Google begins transferring your contacts, phone settings, and apps to your new device.

Other data (such as music, photos, and videos) won't be transferred in this manner from the old Android device to the new one. You need one of many paid apps from the Android Market to achieve this.

Transferring data between BlackBerry devices

The BlackBerry Desktop Manager enables easy transfer of data from an old BlackBerry to a new BlackBerry. Follow these steps:

1. **Connect the old BlackBerry device to the computer.**

2. **Start the BlackBerry Desktop Manager. On the application, select the Device Switch option.**

3. **Select the option to Switch BlackBerry devices.**

   The following screen should show three columns — one for the old device, the next for the new device, and the third for options to sync between the two devices.

4. **In the Options column, select the appropriate data to transfer between the two devices.**

   You can choose to transfer all data or specify the data that you want synced between the two devices.

5. **Select Next to proceed.**

   When the data from the source device is backed up, the application prompts you to connect the BlackBerry device to which you want to transfer the contents.

6. **Connect the new BlackBerry device to proceed.**

   This procedure enables you to transfer not only data such as SMS messages, contacts, and call logs, but also apps from one BlackBerry to another.

Transferring data between Nokia Symbian devices

Nokia Symbian devices have a program called Switch that enables you to transfer data and apps from one device to another, or keep two Symbian devices in sync.

Follow these steps to transfer data from an old Symbian device to a new one:

1. **Start the Switch option by choosing Menu⇨Settings⇨Connectivity⇨ Data Transfer⇨Switch.**

2. **Set a connection type to the appropriate method.**

   For example, if both devices support Bluetooth, select it as the method to connect them.

3. **Connect the two devices by entering a code on both devices.**

   Entering the same code on both devices allows them to be paired.

4. **On the new Symbian device, select the content to be transferred from the other device.**

   The transfer of data proceeds, transferring the information you selected from the old device to the new one.

Exploring Corporate Solutions for Backup and Restore

The options for backing up and restoring data vary from one device to another. Most OS and device vendors offer software applications to facilitate the backup and restore features on their own devices. For example, the iTunes software facilitates backup and restore only for iPhones and iPads. Likewise, the Nokia PC Suite can back up and restore data only for Nokia Symbian smartphones. It's the same story with the BlackBerry Desktop Manager, which works only for BlackBerry devices.

If you're looking to deploy backup and restore for your employees, you're probably alarmed at the prospect of managing so many different applications. Gone are the days when everyone had a BlackBerry smartphone and all you needed to do was install and maintain the BlackBerry Enterprise Server. Now, people use a wide variety of devices, including the popular iPhones and iPads and the increasingly popular Android devices.

It's clearly an administrative nightmare to handle each device with its own unique software application. What you need is an application that can back up data from different types of devices and restore to devices of all those types as well.

There are solutions available from vendors to manage mobile device policies, including backup and restore for all leading smartphone types. They're generally called mobile device management (MDM) solutions. Chapter 15 offers a detailed discussion of solutions for mobile security and device management.

Table 12-1 provides a checklist of things to look for when you're short-listing MDM solutions for deploying backup and restore to your company users.

Table 12-1	Backup and Restore in MDM Solutions
What You Should Ask For	*What to Look For*
Which mobile platforms does the MDM solution support?	Obviously, the more platforms, the better. The platforms that must be supported include the Apple iOS, Android, BlackBerry, and Nokia Symbian.
What types of data can the MDM solution back up?	Typically, you'll want to back up SMS, call logs, phone settings, bookmarks, apps, and user files.
Are there differences in how the solution handles devices from one mobile platform to another?	Often, the features supported on one platform don't apply to every other platform. Be sure you understand what can be backed up, and on what platform.
Is it possible to restore from one type of device to another type?	It will help you a lot if the solution you deploy can restore from (say) an iPhone *to* an Android smartphone, or vice versa.
Where is the data stored?	Because you're looking to back up users' mobile data and apps, it's important to understand where the data will be stored. If the MDM product uses a cloud solution, be sure to understand its security and privacy models.
How long is the data stored?	Be sure you understand how long the data will be stored. Many solutions offer a time-constrained backup. For example, backups retained longer than a certain time period (such as two years) may be purged automatically.

Case Study: AcmeGizmo Backup and Restore Use Cases

Over time, Ivan, the administrator, had become quite comfortable with the mobile device security solutions that he was responsible for rolling out across AcmeGizmo. One afternoon, the deployment was really put to the test. Pete, the President and CEO of AcmeGizmo, had apparently lost his Windows Mobile smartphone somewhere between his home and the airport.

Prior to the deployment of the current security and device-management products, Pete might have risked losing all his data when he misplaced his smartphone. As an end user, his primary option for backing up data from the device would have been to synchronize the device to his PC using the Microsoft Zune desktop software. Unfortunately, it had been over a month since Pete had done so — busy executives often don't consider smartphone backup a day-by-day top priority. Ivan would later inform Pete that he could set his Zune device to back up to his PC automatically, but Pete hadn't done that prior to this loss. The damage was done.

Luckily, Ivan had implemented backup and restore functionality for AcmeGizmo, and Pete's phone was covered. Although this software didn't back up Pete's applications, the core of the most important data on the device had been backed up.

The first thing Ivan did was to remotely wipe the lost device, ensuring that the sensitive corporate data on the device wouldn't end up in the wrong hands. Having settled that issue, Ivan remembered that he had a new Android smartphone that he had been using for internal testing back in his office. Knowing that it was critical for Pete to get up and running quickly, he offered Pete the Android device.

Because the backup and restore solution that Ivan had chosen for AcmeGizmo was able to store device settings in a device-independent fashion, Ivan could take some of the backed-up files from Pete's Windows device and restore them directly to the Android device. Eventually, Pete ordered a new Windows device and could restore everything — with the exception of applications, which Windows doesn't back up.

It could have been so much worse.

Chapter 13

Securing Mobile Applications

· ·

· ·

*T*housands of applications arc available in the marketplace or app stores for mobile devices, including iPhones, iPads, and Android devices. Whether you're downloading apps from the market or deploying apps to your company's employees for their devices, it's important to know what security exists within these applications to protect them from other applications. Mobile operating systems like Apple iOS and Google Android have built-in security infrastructure protecting applications from each other.

Protecting or shielding one application from another is often called *sandboxing*. In effect, the application has its own little sandbox to play in and has no access to data or information stored in other applications' sandboxes.

This chapter examines what sandboxing means for the corporate environment when employees bring their personal phones and tablets (and whatever's on them) into the workplace. It's desirable for the enterprise to deploy security in a way that shields the applications on its employees' mobile devices from one another. The idea is to prevent accidental leakage of corporate information or even theft by a malicious application.

Understanding the Importance of a Sandbox

A *sandbox* is an environment in which each application on a mobile device is allowed to store its information, files, and data securely and protected from other applications. The sandbox forms and maintains a private environment of data and information for each app.

Keep in mind that a sandbox does not make the entire device any safer. If a malicious app is downloaded to the device, it can still infect and steal information from the user directly, even if not from another application. For example, if a simple gaming app downloaded from the Internet asks the user for access to GPS information, that's cause for suspicion. Hackers write apps that seek information directly from unassuming users, and some can assume the guise of a simple and uncomplicated purpose (as with a gaming or utility app).

As an analogy, think about the various applications on a PC: browsers, text editors, and other such everyday applications. A sandbox would make each application write its data into a particular folder that would be inaccessible to other applications. Although this would mean that a virus couldn't access users' e-mails or read a folder maintained by their Word applications, the malware could still cause trouble, stealing other device-specific information or prompting users for credentials. Having a sandbox is therefore not a substitute for having antimalware applications.

Here are the main points to keep in mind about sandboxing mobile applications:

- **Apps can be sandboxed.** Apps can be developed to protect their information and data from each other, preventing one app's data from being read by another app.

- **A sandbox does not equal mobile security.** Sandboxing one or more apps means only that those apps are shielded *from each other*. Malicious apps can still be downloaded that seek information from unprotected apps or directly from the user.

- **A sandbox does not protect the entire device.** To protect the entire device, you need a mobile security app that detects and prevents viruses, spam, Trojans, and other mobile threats from invading in the first place. Chapter 10 describes such threats and their prevention in more detail.

All mobile apps maintain a lot of application-specific information, often stored in files on the mobile device file system. Depending on the application type, the information may include personal or corporate information, which in turn might include confidential or sensitive information. With the prevalence of mobile banking, corporate access, and other such use of smartphones, the information stored by each app is valuable. It's therefore imperative to ensure the safety and security of the data stored by each app.

App Security on Various Platforms

Many enterprises develop their own mobile applications for various platforms, including the BlackBerry, Apple iOS, and Google Android. Some also develop applications for Nokia Symbian and Windows Mobile or Phone platforms. If you are an administrator wanting to deploy your own apps to employees, you should be planning for securing the contents of corporate-owned apps, as well as for protecting the contents of such apps from other third-party apps on the device.

As we mention earlier, sandboxing or securing an app's data from other apps is available on several platforms. In the following sections, we take a look at how app security is accomplished on the three leading platforms enterprises deploy: the BlackBerry, Apple iOS, and Google Android.

App security on BlackBerry devices

As an enterprise administrator, you can control what apps can be deployed on your employees' BlackBerry devices. The BlackBerry Enterprise Server (BES) is an example of a leading Mobile Device Management solution for BlackBerry devices that allows the configuration and enforcement of several application security policies for corporate use. Using BES policies, you can specify whether a user can install third-party apps, or determine the privileges that third-party apps can enjoy on the device.

Third-party apps can, in general, access two types of data on a BlackBerry device:

✔ User data, such as e-mail, calendar, and contacts

✔ App data — persistent storage that shares data with other applications

You can control or restrict access to both types of data by using BES policies. If you develop your own apps for corporate-owned BlackBerry devices, you can enable appropriate permissions for your apps.

The BlackBerry also includes a personal firewall feature that restricts the types of connections maintained by an application. When an app tries to establish an internal connection to a corporate server, the device prompts the user to allow or deny that connection. As an administrator, you can choose to allow or deny such connections as a policy. This prevents suspicious apps from breaking into your corporate network and stealing information from internal servers.

Third-party apps can be written to use BlackBerry device APIs for sensitive packages, classes, or methods. Such apps need to be signed by Research in Motion (RIM) before they are allowed to use those APIs. The signing process ensures that the app is tested and verified for authenticity before being granted APIs to use sensitive information.

App sandboxing on Apple iOS devices

Application developers use the sandboxing capability of Apple iOS to protect the integrity of user data and to ensure that their applications don't share data with other apps installed on the user's device. Each app has access to its own files, preferences, and network resources. Recent versions of iOS have also added the capability to encrypt application data so that sensitive data such as usernames, passwords, or credit card numbers can't be accessed easily from the file system.

A sandbox limits the damage that a potential hacker can do to an Apple iOS device, but it *cannot* prevent an attack from happening. Software defects in an application could allow a hacker to cause the app to crash. Although Apple has built robust sandboxing features into the Apple iOS, it's up to the app developers to ensure that their apps are written securely and to prevent hackers from exploiting user data.

When an app is installed on a mobile device, the system creates a unique folder for it, much like you would do on a regular computer. The path to the app's home directory looks like this:

```
/ApplicationRoot/ApplicationID/
```

The ApplicationRoot folder is where all apps are installed. The ApplicationID is a unique name for each app, and distinctly identifies the app to set it apart from other apps. Each app stores user data and configurations within this folder.

Each app's folder is protected and shielded from access by other apps, as shown in Figure 13-1.

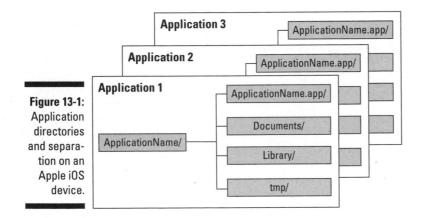

Figure 13-1: Application directories and separation on an Apple iOS device.

Protecting files on iOS devices

On Apple iOS devices, certain files marked by the app developers can even be encrypted when the device is locked. Doing so requires the encryption capability of the device to be enabled and configured. Once that's done, certain types of content can be protected automatically when the device is locked. When the files are locked, not even the app can access their contents.

This feature also extends the protection that shields a particular app's data from another app. Note, however, that this is an optional feature; not all apps need to encrypt files on the file system. A file only gets encrypted if the app developer designates it for automatic protection. Even so, this is a useful feature for app developers, especially if they hold sensitive information on the device (such as the user's username, password, or other credentials).

Sandboxing your apps on Apple iOS devices

If you're in the process of buying apps — whether for your company's employees or for yourself — you'd be well advised to check each app's security capabilities. As noted earlier, some capabilities (such as file encryption) are optional and used at the discretion of the app developer. Therefore, it's worth asking those app developers about the security capabilities of the apps.

If you're considering writing apps for iOS, the native capabilities of iOS allow you to build security within the app itself. For more information about how to develop security within your app, consult the Apple iOS developer documentation.

If you want to deploy corporate apps for your employees' Apple iOS devices, look for Mobile Device Management capabilities that will enable you to set policies governing the use of third-party apps on those devices. Chapter 15 lists the key vendors providing such MDM solutions, many for Apple iOS devices. Be sure to check out those solutions for your specific needs.

Android operating system security

Like the Apple iOS platform, Android has a number of security features built into its operating system. On Android, by default, no application has the permissions needed to perform operations that impact other apps, or, for that matter, the device in general. This arrangement prevents apps from reading information or data stored by other apps, and keeps them from reading the user's personal data (such as contacts and e-mails) stored on the device. This inherent security model forms the basis of the Android operating system.

For one application to share data with any other, it must give explicit permission to the other app to read its data. For example, suppose an app that a user downloads from the Android Market needs permission to know her GPS location — say, an app that shows local restaurants and has to know where the user and her device are in order to do its work. When the user installs such an app, it prompts her for permission to read her GPS location. If she grants it that permission, the app will deliver its GPS-dependent features and won't prompt her for permission later, when she runs it.

Android is based on the Linux Operating System, which has elaborate security mechanisms built in. Each app runs with a distinct system identity (including its Linux User ID and Group name), which is unique for all apps. The Android OS assigns a unique User ID to an app when the app is installed. Linux uses this mechanism to separate apps from each other and protect the system in general.

Even with the security built into the Android OS, users own the ultimate responsibility for protecting their devices. A malicious app in the Android Market could still be written to seek permission to a user's SMS messages, contacts, or GPS location. The security built into the OS protects apps from one another, but does not necessarily shield the user's data from malicious apps. So this security feature is only one piece of the puzzle; it does not preclude the need for mobile security on the Android device.

Be sure to check out the solutions and vendors discussed in Chapter 15. Many of those solutions allow you to configure and enforce corporate compliance policies governing the use of third-party applications on Google Android devices. Depending on your specific needs, one or more of those solutions may be interesting to you.

Exploring Virtualization for Mobile Devices

At a VMWorld conference in 2009, VMWare announced that it was working on building virtualization for mobile devices, which would allow multiple operating systems to run on a single device. In virtualization lingo, this means a *guest VM* would run on a mobile device, separate and potentially different from the *base OS* (the OS that comes with the device). Today, when handset vendors release mobile devices, their products use specific operating systems and are identified as Windows Phone 7 devices, Nokia Symbian devices, Android devices, and so on. For that matter, Apple's devices (such as iPhones and iPads) run the Apple iOS platform as the base OS.

The prospect of running *multiple* operating systems on a single device brings about interesting possibilities. A VM is probably as good as any sandbox could get, completely virtualizing and separating system and device resources from one VM to another. This capability would be especially attractive for work environments; mobile devices could be virtualized in much the same way computers and servers are virtualized presently. Virtualization of servers and desktop computers would allow multiple operating systems (say, Windows and Linux) to run on the same device. Sound too good to be true? Unfortunately, it is — for now.

As of this writing, no mobile devices support virtualization in the manner just described, but vendors such as VMWare have publicly announced their intentions of building such products. Figure 13-2 shows a happy virtual workplace running more than one OS in the form of virtual machines (VMs). Note especially the presence of a *hypervisor,* which is a software module running on the device that allows more than one OS to run on the device at a time. A hypervisor is what enables virtualization of servers as well, allowing them to run multiple operating systems, such as Windows and Mac OS X, in parallel virtual machines.

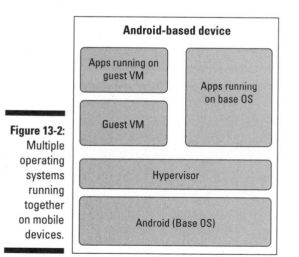

Figure 13-2: Multiple operating systems running together on mobile devices.

Accounting for Personal Devices at Work

Today's workplace is full of high-tech devices that may or may not be under IT control. Many companies find that employees are bringing their personal smartphones and other mobile devices like tablets into the work environment. One way of handling personal devices in a corporate environment is to deploy a corporate sandbox to the device that provides secure application access, protected from other apps that reside on the device. Many enterprises deploy such products that provision a corporate environment separate from the rest of the device, which can host private data and applications belonging to the employee. These products allow secure browsing, e-mail, and application access shielded and protected from other apps on the device.

This separation of personal and corporate footprints on a mobile device is beneficial to employees as well, who would no longer have to carry more than one device around.

Some vendors discussed in Chapter 15 include sandboxing features that allow the mobile users to browse web content, e-mail, and other app data from within a particular vendor's app. The data in these apps is protected and shielded from other third-party apps installed on the device. This is one example of a sandbox solution. When virtualization hits the market for mobile devices, that could be another option, too, enabling you to deploy a virtual machine on an Android device shielded from the rest of the employee's personal device.

Until virtualization becomes a reality for smartphones, be sure to check out some of the vendor solutions discussed in Chapter 15 for your specific sandboxing needs.

Sandboxing Combined with On-Device Security

Running a sandboxing application on your mobile device can be an advantage if the application protects the user data and device from mobile threats. Some apps available in the market provide sandboxing capabilities for particular features such as e-mail. For example, Good Technology (www.good.com) provides application sandboxing for e-mail, which maintains e-mail securely within the application and protects it from access by other applications.

However, as mentioned earlier, sandboxing by itself is not a substitute for real mobile security. You need to complement sandboxing with appropriate mobile security to protect the entire device, and not just the sandbox.

Malicious apps can still attack a device even if one or more apps have sand-boxing implemented. Chapter 9 describes some mobile security threats that a smartphone or tablet device are vulnerable to.

An ideal corporate solution includes such a sandboxing or application security solution combined with an on-device mobile security solution that provides protection from viruses, malware apps, and spam. The sandboxing solution provides application security to your corporate apps as well as to the user's data in private apps downloaded from an app store. The mobile security solution complements this application security by ensuring that files and data received or sent by the device are free of viruses or threats to data or applications.

Be sure to check out the vendors discussed in Chapter 15, many of whom provide application sandboxing or security policy features, or mobile security features, or both.

Part V
The Part of Tens

The 5th Wave
By Rich Tennant

In this part . . .

Where else can you get 20 tidbits of information that are so helpful? Where else but the Part of Tens! Chapters 14 and 15 cover ten of the best places to go online for more info and ten mobile security vendors to help you with your special circumstances, respectively.

If we haven't covered it already, it's here in the Part of Tens.

Chapter 14

Top Ten Online Information Sources

*W*ith this book, we wanted to condense and present all the prevalent threats and solutions as they pertain to mobile devices, but it goes without saying — we'll say it anyway, though — that the types of threats change by the hour.

To secure your mobile devices, you must keep abreast of the latest types of threats and solutions. What better place to go and find this late-breaking information than the Internet? Of course, surfing aimlessly and looking for self-professed experts on this topic may not be a judicious use of your time. Therefore, we compiled this list of trusted online information sources.

Tech SANS

`www.sans.org`

SANS (SysAdmin, Audit, Network, Security) is the precursor to all things security and is a great resource for topical white papers, weekly bulletins and alerts, training, and security vulnerabilities. What makes SANS so unique is that its membership consists of more than 165,000 security professionals, auditors, system administrators, and network administrators that share lessons and solutions to the challenges they face. They've embraced the mobility wave, and you'll find a healthy dose of mobile topics on the Tech SANS website.

Dark Reading

`www.darkreading.com`

The Dark Reading website is devoted to all things security. In-depth security analysis on all aspects of security — including mobile — is its claim to fame. You'll also find frequent webcasts with leading industry luminaries. Tweets, RSS feeds, e-mail . . . there are dozens of ways to keep up with content. The topic of mobility security is starting to appear more frequently on the Dark Reading website, so the good news is that you can keep up with threats. However, the bad news is that this means the threats keep coming.

F-Secure Security Threat Summaries

`www.f-secure.com/en_EMEA-Labs/news-info/threat-summaries`

F-Secure is one of the oldest mobile security companies around. Its threat summaries are comprehensive, yet succinct. Periodically published, they're a great resource to find out about the latest and greatest threats — both as a look back as well as a harbinger of things to come.

Infosecurity Network

`www.infosec.co.uk`

This site is a great collection of blogs, events, and pithy videos on all things security. And the topic of mobile security is a big part of what this site reports on.

National Institute of Standards and Technology (Security Research)

`http://csrc.nist.gov/groups/SNS/index.html`

NIST, an agency of the U.S. Department of Commerce, was founded in 1901 as the nation's first federal physical science research laboratory. Over the years, the scientists and technical staff at NIST have made solid contributions to image processing, DNA diagnostic chips, smoke detectors, and automated error-correcting software for machine tools. NIST sponsors three groups that do cutting-edge security research:

✔ **Cryptographic Technology:** This group develops standards and researches how to keep secure what needs to be secure.

✔ **Systems and Emerging Technologies Security Research:** This group helps to define "emerging" for the rest of us so we can just read the highlights.

✔ **Security Management and Assurance:** This group works with other federal organizations in search of consensus so there are single standards, not multiple ones.

Each group dabbles with mobility, and you'll be ahead of the game if you can find a few moments each day to keep up with the plethora of information published.

Vendors' Websites

```
www.juniper.net/us/en/dm/mobilesecurity/
```

Juniper Networks has state-of-the-art security offerings, and its site is a treasure trove of information for enterprise, mobility, and consumer. The case studies in this book were based on Juniper's security offerings.

```
http://blogs.mcafee.com/mcafee-labs
```

McAfee has great blogs that bring today's security information into focus. This is a good site to see what's happening today in the security world.

```
http://us.trendmicro.com/us/solutions/enterprise/security-
          solutions/endpoint-security/index.html
```

Trend Micro can shed light onto any endpoint security issue you're having — or may have, when your network gets big enough. If you're still confused about the importance of endpoint security, stop here.

```
www.symantec.com/business/theme.jsp?themeid=mobile-
          security-management
```

Symantec is another established name in security and computing. Explore its site to see how the company is constantly adapting its products to a changing world.

ICSA labs

```
www.icsa.net
```

The International Computer Security Association (ICSA) has a website with lots of white papers, pointers, and other useful information for anyone who handles IT security issues. Even though the site isn't targeted specifically at mobile-related issues, it provides broad coverage of a wide range of security issues, some of which are mobile related.

CERT

```
www.cert.org
```

The CERT program is part of the Software Engineering Institute's CERT Coordination Center, a federally funded research and development center at Carnegie Mellon University. Primarily focused on responding to major security incidents and analyzing product vulnerabilities, it has also embraced development and promotion of the usage of appropriate technology and systems management practices to resist attacks on networked systems, to limit damage, and to ensure continuity of critical services.

US-CERT

```
www.us-cert.gov
```

US-CERT is the United States Computer Emergency Readiness Team and coordinates between the government and the public against both small and massive cyber attacks. This site provides important information about the latest threats and vulnerabilities. You can find a good white paper on mobile security at www.us-cert.gov/reading_room/TIP10-105-01.pdf.

GSM Association

```
http://gsmworld.com/our-work/programmes-and-initiatives/
                 fraud-and-security/index.htm
```

The GSMA represents the interests of the worldwide mobile communications industry. Spanning 219 countries, the GSMA unites nearly 800 of the world's mobile operators as well as more than 200 companies in the broader mobile ecosystem, including handset makers, software companies, equipment providers, Internet companies, and media and entertainment organizations. Security has become a topic *du jour* of late, and the website covers some good mobile-specific security topics.

Chapter 15

Top Ten Mobile Security Vendors

*I*f you've followed along so far, we've explored the nuts and bolts of what makes up mobile device security. Now we look at some leading solutions available in the market. These are potential candidates for you to research and consider deploying in your organization.

The mobile device landscape is evolving rapidly, thanks to new devices hitting the market virtually every week. Therefore the vendor solutions we describe in this chapter are expected to evolve just as rapidly to keep in sync with the latest market trends. Be sure to research each solution from the corresponding vendor's website and, ideally, follow up with a trial of the software.

The solutions we describe in this chapter are in alphabetical order, and not necessarily in order of merit or authors' recommendation.

AirWatch

`www.air-watch.com`

AirWatch has a broad solution spanning the Apple iOS, Android, Windows Mobile, BlackBerry, and Symbian platforms. Its solution covers a broad array of mobile device management features. If you're looking for a centralized management solution to configure and deploy policies for all five of these device platforms, you might want to take a look at AirWatch.

Good Technology

www.good.com

Good Technology offers Good Mobile Control, a mobile device management solution for devices running the Apple iOS, Android, Windows Mobile, Nokia Symbian, and HP Palm OS platforms. It offers configuration management, loss and theft protection, and password policy features, among others. Good is reputed to be in the mobile device management space for a long time.

Juniper Networks

www.juniper.net/pulse

Juniper Networks is a leading vendor of networking solutions, including a broad range of routing, switching, and security products and services. Its Junos Pulse Mobile Security Suite includes a combination of its market-leading SSL VPN capabilities combined with mobile security and device management features. The Junos Pulse Mobile Security Suite clients are available for Apple iOS, Android, Windows Mobile, Nokia Symbian, and BlackBerry platforms. Along with mobile device management features, Juniper's solution provides security features, such as antivirus, antispam, and a personal firewall, as well as SSL VPN integrated in the same mobile clients.

Mobile Active Defense

www.mobileactivedefense.com

Mobile Active Defense's Mobile Enterprise Compliance and Security (MECS) product provides a good choice of mobile device management features for Apple iOS, Windows Mobile, Nokia Symbian, and Android devices. Its solution enables the management of mobile device inventory in an enterprise. The MECS solution also includes smartphone firewall capability and policy management features.

McAfee

```
www.mcafee.com
```

McAfee has long provided security antivirus and antispyware software for Windows PCs. McAfee now offers security software for mobile devices through McAfee Enterprise Mobility Management. Its mobile device management and security software is available for Apple iOS, Windows Mobile, and HP WebOS, with limited support for Android.

MobileIron

```
www.mobileiron.com
```

MobileIron's Virtual Smartphone Platform provides mobile device management for Apple iOS, Android, BlackBerry, Windows Mobile, Windows Phone 7, and Nokia Symbian devices. This product provides remote provisioning, mobile device management, and deployment capabilities to manage mobility policies for a broad range of platforms.

Sybase

```
www.sybase.com
```

Sybase's Afaria provides a rich suite of mobile device management features for Apple iOS, Android, BlackBerry, Windows Mobile, HP Palm OS, and Nokia Symbian platforms. The Afaria product includes a number of mobile device management and provisioning features that allow an enterprise to manage policies and applications for mobile devices.

Symantec

```
www.symantec.com
```

Symantec's Mobile Management solution includes security features (antivirus, antispam, and firewall) for Windows Mobile and device management features for Apple iOS, BlackBerry, Android, Nokia Symbian, and Windows Mobile devices. Symantec's solution extends its security features, such as antivirus and firewall, from Windows to mobile platforms.

Tangoe

www.tangoe.com

Tangoe's Mobile Device Management product provides a good breadth of mobile device management features for Apple iOS, Android, BlackBerry, and Windows Mobile devices. Its solution also includes device monitoring features and the ability to control applications installed on corporate smartphones.

Zenprise

www.zenprise.com

Zenprise provides Zenprise MobileManager, an end-to-end lifecycle management solution for BlackBerry, Apple iOS, Android, Windows Mobile, and HP Palm devices. Zenprise's solution includes expense plan management, service plan management, and infrastructure monitoring services in addition to mobile device management features, such as policy enforcement and loss and theft protection.

Getting information from the experts

To deploy an effective mobile security and device management solution, be sure to read what the experts and analysts are saying, too. Gartner (www.gartner.com) publishes regular reports on mobile device management and security vendors. Their reports are useful for analyzing the various offerings in the market.

Index